Other books by Richard Manning

*Land Stand*
*A Good House*
*Grassland*
*One Round River*

# Food's Frontier

# Food's Frontier

## The Next Green Revolution

# Richard Manning

**North Point Press**

A division of Farrar, Straus and Giroux

New York

North Point Press
A division of Farrar, Straus and Giroux
19 Union Square West, New York 10003


Copyright © 2000 by Richard Manning
All rights reserved
Distributed in Canada by Douglas & McIntyre Ltd.
Printed in the United States of America
Designed by Jonathan D. Lippincott
First edition, 2000


Library of Congress Cataloging-in-Publication Data
Manning, Richard, 1951–
    Food's frontier : the next green revolution / Richard Manning.—1st ed.
        p.   cm.
    Includes bibliographical references.
    ISBN 0-86547-593-8 (alk. paper)
    1. Agricultural innovations.  2. Green Revolution.  3. Food supply.  I. Title.
S494.5.I5 M365 2000
338.1′6—dc21                                                            00-30533

For the people doing the work

# Contents

# Food's Frontier

# The Seed

## The Case for a Second Green Revolution

Conditions of life are forcing a fundamental shift in the nature of agriculture. A generation ago, some advances—largely in plant breeding—produced a quantum leap in food production, especially in the developing world, averting worldwide famine. We feed ourselves largely on those earlier gains, which we call the Green Revolution. Now we are in need of another such leap, but we lack the technology to effect it. This is the dilemma that frames all the efforts and debates this book will follow.

A forecast of famine is nothing new. Since the time of Thomas Malthus, who gave his name to the whole enterprise of considering doom, supposedly ironclad calculations have demonstrated the onset of worldwide famine—and many times, they have ultimately proved incorrect. The most famous of these in modern times came from the respected biologist Paul Ehrlich, who in 1969 forecast a starving Japan and China invading Russia in a food war within a decade. Driving his conclusion was the seemingly inexorable collision course of two graph lines: it seemed certain that humanity would outbreed any possible gains in agricultural production. Ehrlich's scenario did not foresee a remarkable blip in one of those lines—the Green Revolution.

Ehrlich didn't know about Norman Borlaug's work, by then already twenty-five years old. Backed by the Rockefeller Foundation, Borlaug had been working in Mexico to breed high-yielding strains of wheat resistant to fungus and rust diseases. The experiment worked so well that wheat heads became top-heavy with fat kernels and toppled over, a problem called "lodging." More breeding, this time with a dwarf Japanese variety to prevent lodging. More breeding, to improve nitrogen use. More tweaking of fertilizers, and so on and so on with this wheat, with parallel work in rice and corn. ("Corn" is the Old English word for grain in general, now applied in the United States to what is more properly called maize. I use the terms "corn" and "maize" interchangeably.) Wheat, rice, and maize are the big three, the trio of grasses that were domesticated in separate parts of the world—rice in south China, wheat in the Middle East and Southwest Asia, maize in central Mexico—and now provide more than half the energy humans consume, in the dense, storable package of carbohydrates that is grain. The net result of the work Borlaug inspired was an average annual increase in harvests of 2.1 percent a year between 1950 and 1990, the compounding growth curve that led to a tripling of harvests during that period. The collision Ehrlich had predicted was avoided.

Asia, the predicted "basket case," became a bread basket that now mostly feeds itself, as does much of Latin America. Periodic pockets of famine remain, but by and large, the world is less hungry than it was a generation ago. When Ehrlich wrote *The Population Bomb*, 56 percent of the world's population lived in nations that provided less than 2,200 calories of food per day per person, a subsistence diet. In its 1992–94 report, the UN's Food and Agricultural Organization estimated that number had dropped to 10 percent. Indeed, a revolution. Hungry people still exist in the world, but proportionately fewer than a generation ago. So why not ride the happy results of this into the future and call the problem solved?

A fitting person to answer this might be Timothy Reeves, director general of CIMMYT, which is the acronym in Spanish for the

International Center for the Improvement of Wheat and Maize. CIMMYT evolved like modern wheat from Borlaug's work and, together with the IRRI, the International Rice Research Institute based in the Philippines, institutionalized the Green Revolution. Borlaug himself still maintains an apartment and offices at CIMMYT's headquarters just outside Mexico City.

Reeves, an Australian, articulate, confident, and direct, sits in his campus-like office building amid the experimental fields of wheat and looks across the brown haze toward Mexico City a half hour away. He's just back in Mexico after visiting state-of-the-art farmers in Nebraska and Australia.

"The thing that really alarms me is I feel most people have underestimated the task," he says. The task he refers to has a simple number attached to it: population growth will double demand for food as soon as the year 2020, by some estimates.

"There's an additional one of those Mexico Cities being added every twelve weeks," he says. "If you tell farmers in Australia or Nebraska that they have to double production in twenty years, they're stopped in their tracks, because . . . all they know is it's going to take new technology, but they can't think about what it would be . . . That gives you some idea of what has to be done in developing countries, if the cutting edge has no idea of what needs to be done."

From the beginning, the Green Revolution has had its critics, especially those who have suggested that its heavy reliance on high inputs of water, capital, and chemical fertilizers and pesticides are simply not sustainable. Reeves himself voices the critics' chief concern: "In feeding ourselves, are we starving our descendants?"

The sense of discomfort with the Green Revolution is no longer limited to its critics. There is consensus that the techniques that have brought us this far will not be able to sustain us in the future. Production is leveling off. Since 1989–90, world grain harvests have risen on average only .5 percent a year, a quarter of the rate of the Green Revolution boom years. Changed political circumstances, particularly the collapse of the Soviet Union and the resulting economic chaos in one of the world's most important grain-producing regions, offer partial explanations, but there are signs that, politics aside, Green Revolution techniques are approaching the limits of what they can produce.

If that's true, not only will supply be constricted but the demand side of the equation will also be thrown into flux. From the beginning, agriculture has been the primary engine of human population growth; the dense package of storable carbohydrates that grains provide allows mobility, cities, hierarchy, technology, medicine, longevity. We count on more agriculture to provide food for ever-growing numbers of people, the solution to the population problem. We forget that the relationship is circular, dynamic, and not at all simple.

Overall, a veneer of good news shines on the population front, what demographers sometimes call a "reproductive revolution," a mirror image of the Green Revolution that has given the planet some respite from the population bomb. What they mean is that, for a variety of reasons like birth control and increased prosperity, fertility worldwide has dropped to an overall annual growth rate of 1.5 percent now, compared to 2 percent in the 1960s. Indeed, in much of the developed world, especially Europe, population growth has stabilized.

This overarching trend, however, masks some problems embedded in the numbers. First, as with much that happens in the world, the trend is geographically lopsided. Large parts of the developing world, precisely the areas least able to grow their own food, still have high birth rates. Even with Green Revolution gains, regions of Africa have more than offset increased crop yields with increased population. Food production per person actually decreased in thirty-one of forty-six African countries in the decade beginning in 1985.

Probably these statistics mask an even greater food crisis, in that an undeniable effect of the Green Revolution has been to displace rural people through mechanization and larger-scale, capital-intensive farms. This occurred in both the developed and less developed world, but in the latter the people displaced were often subsistence farmers. Their produce often doesn't show up in yield statistics, but it used to feed people. Displaced to the cities, this class of people no longer feed themselves.

Meanwhile, there is a well-established correlation between an increase in income and declining birth rates, to the point that development accounts for much of the reproductive revolution. But there is also a correlation between increased income and consumption of meat, which in turn greatly ratchets up the demand on grain. (It takes about seven grams of grain to make a gram of beef.) Forecasters expect demand for grain for human food to increase by 47 percent in the developing world by the year 2020. At the same time, demand for grain for livestock is forecast to jump 101 percent during the same period.

Finally, the current low birth rate is only one factor determining population growth. Another factor is the bulge of people of reproductive age who were themselves the result of the earlier boom. A lower birth rate applied to a higher base still yields a lot of new mouths. UN projections say there will be 8 billion humans by 2025. This is what drives the sense of urgency among agronomists and agricultural economists.

According to projections by the International Food Policy Research Institute, there will be 150 million malnourished children under the age of six among us in the year 2020. That is a decline from the present percentage, but one out of four children on the planet would still be malnourished, with the heaviest concentration in South Asia and Africa. These projections are based on an agriculture that continues along the curves carved by the Green Revolution, an assumption perhaps more responsible for the uneasiness among experts than the raw numbers of people.


Begin by considering the United States, which represents the cutting edge of agricultural productivity. Average grain yields in 1960 were

Figure 1. Chronic Undernutrition

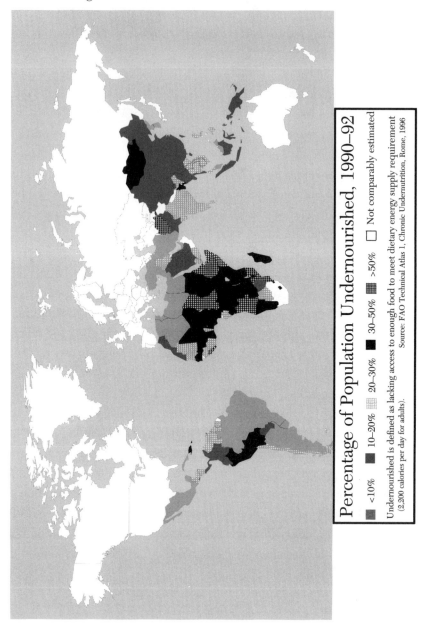

Percentage of Population Undernourished, 1990–92

■ <10%  ■ 10–20%  ■ 20–30%  ■ 30–50%  ▦ >50%  □ Not comparably estimated

Undernourished is defined as lacking access to enough food to meet dietary energy supply requirement
(2,200 calories per day for adults).        Source: FAO Technical Atlas 1, Chronic Undernutrition, Rome, 1996

45 percent higher than in 1950. During the decade 1960–70, the increase was 43 percent, then 20 percent during the next decade, and 10 percent during the next.

During the same period, the United States and the rest of the world increased harvests by boosting yield per acre, and also by bringing more land under production. At present there is no more uncultivated land to be converted to crops. In some areas, the supply of arable land is decreasing, especially in the United States, where urbanization continues to chew up farms. In the 1970s, the United States brought into cultivation about 50 million hectares (a hectare is 2.5 acres) of land considered "highly erodable" and was forced to retire it a decade later, after it was found to be losing more than thirty tons of topsoil per year per hectare to erosion. The land was seeded back to grass. Furthermore, much of the yield increase came about by increasing irrigation, yet supplies of fresh water are rapidly being exhausted. Much of the United States' prime grain lands overlay the Ogallala Aquifer that lies under parts of Nebraska, Oklahoma, Texas, New Mexico, Kansas, South Dakota, and Colorado. This fossil water aquifer has been pumped to depletion in some irrigated areas, and will be entirely depleted at current rates in a few decades.

Much of the gain in yields came through increased use of fertilizer, yet fertilizer application rates have reached their maximum in the developed world, the point of diminishing returns. Any more fertilizer simply couldn't be absorbed by the crops and therefore would not bring a corresponding increase in yields. Farmers know this and are already starting to cut back.

Gain also came through improved crop varieties, yet this strategy, too, has an upper limit. Increasing yield is really a matter of coaxing plants through selection to adopt a strategy that favors seed over all other plant parts: roots, leaves, and stems. Agronomists measure the improvement with something called the harvest index, the weight of grain a plant produces compared to the weight of everything else. At the beginning of this century, grain crops typically had harvest indices of .25, a quarter of a plant's total bulk. The Green Revolution's plant breeders brought the index to about .50. There is some more yield to be gained this way, researchers think, but they also believe the maximum index possible is about .60. Beyond that, in a manner of speaking, there is no blood in the turnip.

In July 1998 reports of famine in Sudan became widespread, as did reports of the collapse of grain prices in the United States, a fall dominoed off the collapse of the Asian economy and its markets for grain. The Clinton administration quickly announced a major purchase of grain for famine relief to Sudan. Arguably it was sheer coincidence that the United States was but one month later firing missiles at Sudan, but nonetheless a somewhat dramatic illustration that food aid is more related to propping up U.S. grain prices than to genuine aid, and that it more often than not precipitates unrest. Aside from those selling the grain, few engaged in international food issues believe massive food aid—some call it dumping—does very much more than torpedo prices and undermine farmers' income in the countries receiving the aid.

"The biggest disincentive [to promoting food production] we have seen has been the dumping of low-cost grain," says CIMMYT's Timothy Reeves. "It tears the guts out of farmers, and it takes all the political resolve out of the government that they might have for farming."

Nor is helping simply a matter of good intentions. Take an example from Mozambique in the mid-nineties, typical of a phenomenon repeated through the Green Revolution. Researchers attacked hunger in Mozambique not by dumping grain but with fertilizer and improved maize varieties that led to a fivefold increase over traditional yields. *Scientific American* reported: "The poor conditions of local roads prevented farmers from transporting their produce. The northern area (of Mozambique) ended up awash in maize, with stockpiles rotting, and prices fell to a ruinous $40 per ton."

Failure can be even more finely tuned.

John Axtell is a genial man, a professor of agronomy at Indiana's Purdue University. From this base deep in the U.S. corn belt, he has spent a career tinkering with the mechanics of the Green Revolution, especially with sorghum in North and East Africa. Sorghum, now an important cereal crop worldwide, originated in the center of agriculture in Africa along a long, dry arc that extends from Niger and Sudan in the northeast through Ethiopia and south along the east edge of the continent. From this region Axtell has gathered a series of anecdotes to school some humility into students learning about what is known about food security—the effort to ensure an adequate long-term global food supply.

The Western agronomists who went to help these farmers found them using antiquated threshing techniques that involved spreading the grain on the ground on a pad and walking livestock over it, a problem fixed simply by replacing cow hooves with modern combines. Except that these people, who used traditional methods, did not suffer from iron deficiencies, while those eating a similar diet in India did. Their "old-fashioned" threshing methods were adding enough iron from the soil to compensate for the deficiency in the sorghum.

Through much of this region, farmers grow a dark sorghum, largely, researchers once believed, because they don't know about the more nutritious varieties of lighter sorghum. Sorghum's shade of brown deepens according to the amount of tannin it contains, and tannin, in addition to being bitter, blocks the human digestive system from processing the protein naturally available in the sorghum. Lab animals fed a diet of dark sorghum will waste away. The researchers urged farmers to adopt the light varieties.

There is, however, a common bird in the sorghum belt of Africa, the quelea, that likes sorghum and hates tannin, so the light varieties of sorghum hauled in by well-meaning developed-world agronomists went to the birds. Which is why the farmers of the region grow brown, high-tannin sorghum. The practice co-evolved with birds as a defense mechanism. Axtell says the final piece of the sorghum puzzle fell into place when a student at Purdue whose father was a chief in Uganda asked villagers about the processing of sorghum. He found they always soaked brown sorghum in wood ash before cooking it. The ashes removed the tannin. This bit of information led to a process that now helps U.S. growers raise high-tannin sorghums.

The base of Axtell's respect for the traditional wisdom of farmers was built in 1973, when a survey of 20,000 varieties of sorghum from throughout the world identified two with a very high protein content. The two varieties had been collected by the Rockefeller Foundation from the same area of Ethiopia, so Axtell went there and asked around. Everyone knew exactly what he was talking about. "Kids there know their cereal varieties the way our kids know cars and baseball," he says. One of the varieties was called *wetet begunche*, or "milk in my mouth." The other's name translated as "honey squirts out of it," because it was eaten as a sort of snack or treat, but an important one. The latter variety was generally the first harvested at the end of the dry season, the hungry season, and was roasted for a shot of protein just when hungry people needed it most. The names indicate that people somehow knew the protein was there.

The lesson Axtell takes from these stories is that securing a food supply rests on more than just boosting yield. Both varieties cited here produced less than others, yet local farmers preserved these sorghum strains, even when higher-yielding varieties were available. Survival in a difficult place demanded it.

The Green Revolution at its most fundamental level treated all the world the same, but the lessons being learned in agriculture now are local. A practice, a variety, a people, and a crop endure in a place because selection has finely tuned them to survival. They have evolved along with local conditions, and the path to a sustainable future requires some respect for the results of that process. Food security debates often began, as this one did, by pointing to Paul Ehrlich's failed prediction, but too often we forget that Ehrlich, along with Peter Raven, hatched the concept of co-evolution, an idea that will have much to say about where we go from here.

It is no stretch to say that Don Duvick owes his comfortable suburban home on the outskirts of Des Moines, Iowa, to the Green Revolution. He is a retired senior vice president for research at Pioneer Hi-Bred International, Inc., a lifelong plant breeder with a giant seed company pivotal in the development of industrial agriculture. He is also an affiliate professor of plant breeding at Iowa State University. But we can also understand some deeper layers of his loyalty to agribusiness by noting that the acre or so in front of his house—the spot usually devoted to clipped bluegrass lawns in these reaches of middle America—is given over to a restored tallgrass prairie, the ecosystem that row-crop agriculture eradicated here in Iowa.

Duvick speaks more like a professor than a retired executive. His conversation is thoughtful, measured with lots of alternatives modulating the even flow of his ideas. Like many in agribusiness in his generation, Duvick began life on a small Midwestern farm, an experience that no doubt laid some of the foundation for a journal article he published in 1995:

> Small farms and small towns are not necessarily a superior source of societal and environmental virtues. Small farmers supported by small towns despoiled their own countryside in the 18th and 19th centuries. The farms were then abandoned, and the despoilers went west to states such as Wisconsin where the cycle was repeated, inspiring Aldo Leopold to his monumental work on behalf of the environment. Small farmers supported by small towns in the Great Plains despoiled the grasslands of Oklahoma and Kansas in the late 19th and early 20th century, giving rise to the famous Dust Bowl in the drought years of the early 1930s. In our own time, I

read every day in my local newspaper of the crimes that abound in small towns and on small farms in Iowa. They are as ferocious and as perverted as in any metropolis.

This recounting does not prove that small farms and small towns are worse than the rest of the nation in civic virtue or reverence for the environment. Nor does it say that large farms or large business firms are likely to be superior in these regards. It merely says that no group has a monopoly on virtue or vice, on wisdom or folly, on generosity or greed. To assume otherwise, to assign a class uniformity where one does not exist, will put needless roadblocks on the path to the solution of the real problems facing U.S. agriculture. We must look for solutions where we can find them.

Duvick's pragmatism is deep-seated and channeled by his particular small-farm experience in the 1930s. It is at the same time typical of a whole generation of people his age whose careers played out in agribusiness. That is, they grew up on tough, hand-to-mouth farms when a single event brought some prosperity, enough to send them off to a land-grant school someplace for an education that gave them a shot at a white-collar job. That single event, he says, was the coming of hybrid seed. It transformed Duvick's own life and those of many like him.

The term for the magic of hybrids is "heterosis," hybrid vigor. Hybrids, which are sterile crosses of plants that do not normally cross, yield far more than either parent, the grip on the bootstraps that has boosted most of our yield gains in corn during the past thirty years. Wheat does not readily lend itself to hybridization, but plant breeding, conventional crosses, has created a parallel process. Hybrid or otherwise, new varieties have fueled the revolution.

Duvick, who sits on the board of the International Rice Research Institute, says the early high-yielding rice varieties produced about ten tons per hectare, and subsequent varieties have raised even that bar. But those same early varieties grown today yield only about seven tons per hectare. Cotton breeders report a similar phenomenon. "People are really scratching their heads, saying, Do we have to run this fast in order to stand still?" says Duvick.

He believes that growing conditions are deteriorating, probably because of microbial reactions in the soil, that resistance to disease is declining, and insect damage is on the rise. So far, breeding has more than offset the resulting losses. But breeding has its limits.

For instance, an agronomist in Nebraska has researched corn contest records for fifty years and found that there has been no real increase in the prizewinning yields, which represent a sort of theoretical maximum. True, average yields have increased, but the overall increase has been the result of latecomers, farmers finally adopting what is essentially old technology. The yield has a ceiling.

During this breeder's career, the critical past thirty years, corn yields have tripled. So can the next generation of breeders match that performance to meet projected increases in demand?

"I don't think so," Duvick says.

The more we chase our basic human desire to understand how the world works, the more we reveal our ignorance. People in agriculture might agree with this but are just as likely to say we are most ignorant about social systems, not science. We think we have the technical ability to deal with the problems of producing food, but we fail because of lack of knowledge about such areas as markets and government. True enough; our ignorance in these areas is vast, yet it is equally formidable concerning the raw science—for instance, knowing the workings of a single square foot of soil.

Dr. Eugene Kroonin of the National Center for Biotechnology summarized the situation for *The New York Times*: "Five years ago, we were very confident and arrogant in our ignorance. Now we are starting to see the true complexity of life."

There is a glimpse of this complexity in Robert Goodman's lab group at the University of Wisconsin at Madison, where a grad student might buttonhole you and insist that microbes rule the world. She will explain something of the recent work of microbiologists that has redrawn the tree of life.

Humans launched their first forays into the microbial world in the seventeenth century, with Antonie van Leeuwenhoek's invention of the microscope. That instrument, coupled with the capability to culture, grow, and compare colonies of microbes in the lab, produced a flourishing field of knowledge on which much of modern science rests. Like most tools, though, culturing of microbes obscured as much as it illuminated. Microbiology took a look at the universe by sampling it onto petri dishes and assumed incorrectly that what grew was what the world held. New tools, like techniques for sampling and cataloging DNA, are showing us some signatures not seen before. Now we are beginning to understand that culturing revealed something less than 1 percent of microbial life, leaving a rich world hidden in the shadows.

This new view, spelled out by Carl Woese, of the University of Illinois, in the mid-seventies, showed something even more profound, that the old picture missed a major branch on life's tree—that, earlier, life was plants and animals, fungi and bacteria, but now life is eukarya, bacteria, and the new stuff: the frontier, a whole new category of life, the archaea. In this scheme, most of life we know—visible life, plants, animals, and fungi—fall on the branch called eukarya, meaning organisms built from cells with nuclei. Both archaea and bacteria are prokaryotes, without nuclei. The implication of this grouping is that the latter two, co-equal branches on the tree, harbor the same degree of complexity and diversity as the eukarya. The more we come to understand about all this, the more this impression is justified.

Within the archaea there are forms of life as unrelated to each other as humans are removed from mushrooms. Archaea were the side of life missed by culturing. These are, like bacteria, single-celled organisms, but live by managing their DNA differently. Put in simpler terms, they are just plain weird. The initial work found archaea in places like hot pools at Yellowstone National Park and around volcanic vents along ocean floors. These are the thermophiles, or heat-loving archaea, that probably hold the key to the origins of life. At first it was generally believed the archaea were odd forms evolved to occupy the extreme niches, but then people started looking in ordinary places. Goodman's lab found them in soils and more recently in surprising abundance on the roots of plants. It is not known what they are doing there, but Goodman suspects they are mediating the flow of minerals and nutrients, especially nitrogen, warding off pathogens, coupling with other microbes to create the unseen ecosystem that is the basis of agriculture. The archaea may have begun all of life, taking energy from volcanic vents, yet it may be equally true that the archaea continue to sponsor the exchange between elemental and sentient, which is to say, food.

Jeff Dillen is an engaging young man who was working toward a law degree at the University of Wisconsin when I met him. At the same time, he became an accomplished microbiologist just for the mental exercise, the way some students would take up weight lifting in their spare hours. Like all the rest in Goodman's lab, he is animated by a sense of the frontier; the researchers' slides of the bar code–like registers of patterns in DNA are a window to something heretofore unseen. Dillen did an interesting and straightforward thing: he sampled soil from an organic farm and compared it to soil from one using Green Revolution methods. The archaeal microbes were present in the organic soil but not in the samples from the industrial ag land. We don't know what the archaeal bacteria have to do with the health of crops, but this simple result indicates we should.

Goodman's analogy is to the spotted owl, the endangered species of the Northwest rain forest that signals not so much its own peril as the status of the ecosystem it represents and, by extension, the humans that depend absolutely on its health. Can we know what danger our methods pose to an ecosystem when we don't know how it works, or even what is in it, for that matter?

Vern Ruttan is an agricultural economist at the University of Minnesota, a self-described, unreconstructed Green Revolutionary. He says, "There are no revolutions in the offing. We'll be fighting in the trenches. The gains in the future are likely to come harder and be smaller."

Ruttan was at a meeting in Wayzata, Minnesota, an event that became known as the Spring Hill Conference, named for the place that held it. Goodman, Duvick, and Axtell were there as well, along with more than a dozen agricultural researchers from around the globe. In addition to the brain trust, James M. and Patricia S. Binger represented the board of directors of the McKnight Foundation, along with executive vice president Michael O'Keefe. William M. McKnight, then chairman of the 3M Company, set up the foundation in 1953. Forty-five years later, it was giving away $76 million annually to a variety of causes, one of which is securing international food supply. McKnight organized the conference in Wayzata spearheaded by O'Keefe and Goodman. The idea was to use McKnight's money as a lever and the gathered expertise as a fulcrum against the looming problem. But money and brains had been pitted against this problem before, much larger sums of money than a small Midwestern foundation could muster. Primarily, the conference was looking for a place to begin, a crack in the face of the problem that might yield a purchase point for the lever.

The conference decided to take advantage of some factors already in place. World food aid is not a new thing. Since World War II, the United States has been actively involved, first by attempting a direct export of technology, which largely failed because much of it was inappropriate for the developing world. Then the United States exported institutions to mirror the extension service, a network connecting farmers to expertise at U.S. land-grant universities. This also failed to take root. More recent efforts have centered on building intellectual capacity within developing countries. This last phase has taken a generation, but it has shown results. There are trained agricultural scientists throughout the developing world now—assets in place.

Clearly, the developing world is where the gains are to be made. That's where yields lag, despite a range of available technologies; where the effort is frustrated by a variety of failures of social institutions. Any gains to be made now will have to face social questions. These factors point to a position for the lever, but only point. Precise location can be identified only locally, within the specific context of specific developing-world nations.

The conference set up a way to do that. A plan evolved to award a series of grants to specific projects in the developing world. McKnight committed $12 million over six years. The specifics of the projects would be hatched by each nation's scientists identifying their particular nation's needs, but each program would be required to form a partnership with an American university and American scientists in order to provide the scientific firepower of the developed world, and also to provide some further training for national scientists. Just as important, the partnerships were to open a reciprocal flow of information such as Axtell outlined above in the case of sorghum.

Finally, it was clear that any program would have to be interdisciplinary to bring a broad range of ideas to bear on the broad base of the problem. It would involve not only agronomists and plant breeders but economists and anthropologists, ethnobotanists and even ecologists.

McKnight put out a call for proposals. There was an immediate and overwhelming response from around the world. Goodman had figured on getting about 100 applications. Instead, 450 arrived.

"When the deluge hit, it was quite a shock," he says. McKnight took it as a sign that they had identified a major need.

Goodman had already set up an oversight committee, the skeleton of which came from the Wayzata conference. It included himself as chair; Axtell; Almiro Blumenschein, a Brazilian plant geneticist; Duvick; and Ruttan; and added to that base Sandra Batie of Michigan State University; Jojah Koswara, a professor and director of research for the Directorate General of Higher Education in Indonesia; Molly Kyle Jahn, a professor of plant breeding at Cornell; Alison Power, an ecologist from Cornell; Mandivamba Rukuni, a professor of agricultural economics at the University of Zimbabwe; Usha Vijayraghavan, a microbiology professor at the Indian Institute of Science; and Qifa Zhang, a professor of crop genetic improvement at Huazhong Agricultural University in the People's Republic of China.

The board winnowed the applications to nine projects and began funding them in 1995. Like the problem itself, they are diverse. Goodman acknowledges taking a sort of portfolio concept, balancing some risky, high-tech ventures with some that seemed likely to produce immediate gains.

I owe the reader a disclosure at this point. The McKnight Foundation hired me to write this book and paid for my research expenses. That is a slippery slope for a journalist, made less slippery because this project is run by a group of independent scientists, each with separate biases. For the same reason that they are an asset to the project, they are also assets to this book: they bring to the project some critical intellectual power that will help reveal the story in a variety of lights. At the same time, the committee is a buffer: its independence was extended to include mine. The foundation explicitly granted me editorial control over this project, a license to arrive at and write my own conclusions. I was never denied access to a single meeting, document, or source. I was, in fact, granted access to several meetings where touchy allegations were made and dirty laundry aired. Still, I was never asked to avoid a topic or line of questioning, nor did I hear the phrase "off the record."

Still, McKnight did pay for the research that made this book possible. It is not the sort of work that publishers are eager to underwrite. (In my more cynical moments, I think that it would be far easier to sell a book about a single celebrity missing lunch than the prospect of a quarter of the world's children being malnourished.) The financial basis on some level opens the project to some criticism. Perhaps it will come from independent journalists in the employ of Disney, Time-Warner, or Knight-Ridder.

Now we travel: to Ethiopia, where developed-world science is being applied to a traditional cereal called tef; to Uganda, where sweet-potato research rests on human capacity decimated by conflict; to Zimbabwe, where work on tannins in sorghum is under way: to India, where a range of tricks from traditional plant breeding to genetic engineering is being applied to chickpea improvement; to China, to see if transgenic bugs can prevent a rice virus disease and if wild relatives of wheat can prevent rust; to Brazil and Chile, for an advance look at some new potatoes free of insecticides; to Mexico and the milpa system that fed the Aztecs; and to Peru's highlands, where indigenous tubers are the center of village life.

The program has not solved the world's problems, nor is it the last word in food security. A whole range of institutions, foundations, and organizations is chipping away at the problem. It is, however, a window into what the world is doing about agriculture, and what urgently needs to be done.

# An Island in Africa

## Global Methods, Local Choices

### [Ethiopia]

The road we are on stretches out in two lanes of potholes set in a matrix of what was once probably asphalt. A line of large overloaded trucks and overloaded buses bellows black smoke, bobs, dips, weaves. Donkey carts flutter around traffic like birds among angry beasts. Even well beyond the city, both shoulders of the road are bedecked by an unending row of people, a line of hollow, blank faces set just beyond reach of the passenger, who sits behind a rolled-up window. Right behind the row of people, more people in clots and bunches sell goats and vegetables, catch buses. Behind them, more faces can be seen framed in the shadows of doorways and windows of row on row of tin shacks. Now and again, a new marble building rises a story or two above the shacks. Barefoot men shinny a lash-up of stick scaffolding. Two trucks have collapsed beside the road like battle-spent elephants. An ant line of people scavenges the carcasses for cargo and parts. A fog of diesel smoke fuzzes the whole scene.

It takes us two hours to cover 45 kilometers on this road, one of Ethiopia's principal highways, the vital link from the central and capital city of Addis Ababa south then east to the port of Djibouti, the trunk line that carries coffee and leather to earn the foreign exchange that raises new marble buildings by agriculture's bootstraps. Development depends on the road, yet abstractions like development don't register nearly as sharply as the faces along it. There is no getting used to seeing a young polio victim crawling like a crab over the roadside potholes to beg from passing cars, no getting used to the endless streams of high-boned Ethiopian faces, beautiful young faces of the sort that, in another world, stare from fashion magazines, here staring out from insurmountable poverty.

The road begins in Addis Ababa, where the national museum holds the bones of Lucy, who lived in this region 3.2 million years ago. The most famous of human fossils, she represents perhaps the oldest-known individual in the line of human faces. That doesn't make her the first hominid. Still, Lucy and much of the rest of the fossil evidence suggest that the human endeavor began nearby in Africa's Rift Valley. The road from Addis goes there. If the Rift Valley represents the beginning, does the scene along the road portray the end?

Outside the city the tin shacks thin to a row or two surrounding farm fields. Both shoulders are lined with people, but now they are younger, schoolchildren walking four and five abreast, as in a parade, all in clumps of color, a patch of bright blue, then another bunch all wearing yellow shirts and blouses, school colors flagging the way to class. The line stretches unbroken down the highway to where the crest of the next hill forms a horizon a couple of kilometers away. Behind the line of children on each side of the road, the landscape is tiled with mint-green farm fields, most of them growing a short cereal crop called tef, an odd grain unique to Ethiopia, older than this ancient land's recorded history. Tef has much to do with the road Ethiopia is on and the rows of children that line it.

Debre Zeit, a small city, sprawls 50 kilometers south of the capital in the Rift Valley. On this day in the early fall of 1998, Debre Zeit bustles with militarism: jeeps loaded with camouflaged soldiers, with stern, set faces and rifles at the ready; here and there the white face of a Russian adviser on hand to maintain the aging fleet of MiGs that built the Ethiopian military when its government was Marxist, a period called the Dergue that ended with the deposing of Mengistu Haile Mariam in 1990. The bustle is real now and on everybody's mind throughout the country. Shots have already been fired in a border skirmish with Eritrea. Diplomatic efforts by the United States were holding back the full-scale war that most Ethiopians thought inevitable, and it later turned out they were right.

This war is not a sideshow, but in many of its aspects it is wrapped in Ethiopia's present woeful state, its history of periodic famine, and especially its status as a pawn in global Cold War politics. We may arbitrarily set a beginning point for this modern history at 1930, when a series of succession skirmishes led to the crowning of Haile Selassie as monarch. He presided over what was simply a feudalist society, largely dependent on subsistence agriculture, a nineteenth-century place. The outside world, however, intervened forcefully with the Italian occupation of Ethiopia during World War II, straightforward colonization that deposed Selassie. He weathered out the war in exile, but returned to power at its end with imperial designs. The Allies agreed to these, mostly to allow Ethiopia to create a "federation" with neighboring Eritrea, which had also been occupied by the Italians. Both Ethiopia's and the Allies' motives in this were clear. Eritrea fronted the Red Sea, giving landlocked Ethiopia a port and the West a guaranteed friendly hold on the Red Sea. The federation soon devolved to outright colonization, a grudge that the Eritreans still hold to the current war, despite their final gain of independence from Ethiopia.

The United States especially among Western Bloc countries propped up Ethiopia with copious military aid; half of all U.S. money for arms for Africa went to Ethiopia during the early years of the Cold War. Selassie, however, did little to develop his country, modernize its agriculture, or break up feudalism. Famine resulted, as did political upheaval and finally the execution of Selassie in a coup that was superimposed on the framework of global politics. Overtly, the pendulum swung, when Mengistu seized power in the name of socialism. But for Ethiopians, this mattered hardly at all. He ruled as a despot mostly interested in maintaining his own status. In an early shakedown, for instance, Mengistu consolidated his hold by executing fifty-seven of his own government's officials, including two previous prime ministers. Meanwhile, the country's infrastructure continued to deteriorate, a process accelerated by rampant brain drain. Educated people simply left the terror that surrounded them. The army was killing people by the hundreds to maintain the Dergue's control; periodic famines escalated, especially one in 1979 that affected a million people and killed an estimated 10,000, an event mild in comparison to the much broader famine in 1985. Then, one in five people in the country were affected, and in the northeast, where conditions were the worst, one million people died. That devastation precipitated rampant unrest in the country. The Dergue finally ended in the early 1990s, with the general collapse of socialism worldwide and the decline in aid that had been propping up Mengistu.

At the edge of Debre Zeit, there is a compound, the former country estate of Selassie's family, that has served as Ethiopia's leading agricultural research station since 1955. Its fortunes have paralleled the twists and turns of this country, and it looks it. There's a parklike core that says royalty once relaxed here, but mostly the buildings are stark, institutional, and bare.

During the Cold War and Selassie's period of close ties with Washington and the Eisenhower administration, he visited Oklahoma State University. He became so impressed with the U.S. model of land-grant institutions that he decided to duplicate the system in his own country, thus launching the Debre Zeit Agricultural Research Center.

It was an embryonic attempt but, nonetheless, needed. Ethiopia is at its heart agricultural and has been for millennia. A scan of its landscape today or a scan of its vital statistics would never suggest wealth. Its per capita gross domestic product rests at about $120, among the lowest in the world. Yet in a sense Ethiopia was richly endowed in biodiversity, one of those few spots on the planet where the world's agricultural plants originated. Much of Ethiopia lies isolated on a high, fertile plateau made for agriculture. The surrounding area is lower and drier, a sort of topographical desert in which Ethiopia stands out as an island. The country evolved under its own terms. This physical isolation guards its uniqueness, a condition that can be read today in the distinctive faces of its people and in its language, but also in its crops.

Tef is a key part of this biological legacy, but coffee also originated in the Kaffa province of Ethiopia, a fact still encoded in its name and venerated in the coffee ceremonies ubiquitous in the country. In most hotels as well as in village huts one can watch white-robed women preside over charcoal braziers and beaker-shaped black clay pots in a ceremony redolent with incense and roasting coffee. Fully 60 percent of Ethiopia's foreign exchange comes from coffee exports. The evolutionary creativity of this same plateau also gave us durum wheat, barley, a form of millet, lentils and chickpeas, and an oilseed crop called *noug*.

This rich background may help explain the unmistakable sense of pride in Hailu Tefera's voice, despite the poverty and backwardness all around. He waves an arm across his country's landscape and announces, "This is tef."

Tefera acts as my guide on this road; he and a driver have collected me in Addis Ababa, and now we have worked our way south toward Debre Zeit, far enough to break from the city's solid clot of tin shacks to the broad green plateau punctuated here and there by a hillock studded with huts. Ethiopia is Africa's third-largest country, with a population pushing 60 million, but 50 million of those people are rural, mostly farmers working small plots of about four acres. Tefera says rural Ethiopians prefer to live on high ground, so villages cluster on hillsides with the valleys below kept for small plots.

Most of them, practically every single farmer, grow tef as the staple, as the base. Other grains have a long history of cultivation here, and many farmers grow the rest of the indigenous legacy: sorghum, millet, lentils, and chickpeas. There have been Green Revolution incursions as well. A taste for maize has spread here as it has throughout Africa. Work by the food-aid project sponsored by the international food security group Sasakawa Global 2000 has boosted maize yields and cultivation to the point that once-famine-stricken Ethiopia has become an exporter of maize. Still, Ethiopian agriculture is built on tef. Ethiopian farmers give it more land than any other crop, a total of about 5 million acres.

Its superfine seed grain is ground to a meal and fermented, then made into a pizza-sized flat bread, spongy and full of carbon dioxide bubbles from the fermentation process. Called *injera*, it is to Ethiopia what tortillas are to Mexico. It shows up at every meal. Usually it serves at the same time as a dish, an eating utensil, and food. Typically a piece of *injera* covers the bottom of a wok-shaped bowl. An assortment of stews and sauces are ladled into the center. The diners rip off a hunk of the edge of the *injera*, use it to scoop up stew, and eat the whole business.

Because it looks more like a foam rubber mattress pad than bread, more than one Westerner, on encountering *injera* for the first time, has been known to neatly unroll it across his lap, mistaking it for a thick napkin. On discovering the error and tasting it, more than one has still thought it made a better napkin than food. *Injera* is an acquired taste, but there can be no doubt that Ethiopians have acquired it.

"I myself do not feel satisfied unless I have eaten *injera*," one somewhat dandified upper-crust Ethiopian bureaucrat explained to me, almost apologetically. Expatriates' eyes will gleam at mention of the word. Farmers in the grain belt in the United States have begun planting tef to meet the demand generated by Ethiopian immigrants in American cities. Tef is also known and researched in Israel, which accepted the influx of Felasha Jews from Ethiopia during the

Mengistu years. Yet there have been periodic attempts—especially during the Mengistu years, when tef was frowned upon as backward peasant food—to modernize Ethiopian agriculture with Green Revolution crops.

"They are saying that our people can eat rice," says Tefera. "Forcing our people to eat rice is not an easy thing." Nor a sensible thing. Ethiopia is a cultural and an ecological island in which tef has continuously co-evolved with appetite. Tefera is a plant breeder, with a Ph.D. from the University of London and training in advanced molecular-marking techniques at Texas Tech. All his work has been in tef. He believes that the farmers of his country grow it from far more than habit.

First, come the hell of drought or the high water that periodically leaves fields waterlogged and roots starved for oxygen, tef still produces a crop. Even in the worst years, it yields something. There may be no surplus to sell, but tef provides food to eat throughout the dry season. Other crops may outyield it in both tonnage and value, but tef is still the safest bet and so lies at the heart not just of Ethiopian agriculture but of its food security. When conditions are so severe that there is no grain, the plant still produces its straw, which livestock like. This is not a small factor in a meat-eating country where goats are ubiquitous, even in the busiest sections of downtown Addis Ababa. Goats abound, but vegetation for grazing does not, so a large part of the country's meat supply rests on tef, a second link to food security.

Also, the national appetite for tef, coupled with the fact that there is no competition from imports, means there always is a demand and always a stable, solid price. Ethiopia's island serves its farmers well in this regard also. So where's the problem?

Tefera and I leave his office at Debre Zeit for a walk down neat rows of tended flowers and hedges, down a road full of townspeople on foot and in donkey carts, to visit the sort of grid-set test plots that rim any agricultural research station worldwide. There's a rude hut of brush toward the center of the plots where a guard spends nights shooing off rabbits and antelope. Dozens of women stoop over growing grain, hand-weeding each patch. It is late in the rainy season, and the field is muddy, the tef tall, but arched in a graceful curve like fine, feminine hair. The plots are mostly a liquid, light green, but here and there squares are purple or gray.

Each seed, now nearly fully formed, is the size of a speck of cornmeal, a sand grain of a seed. On the growing plant, seeds are set in florets bundled in threes, and each bundle set at the end of a spikelet ties to the seed head, an effect that gives the grass head or panicle its airy quality. There are as many as 6,000 florets in a panicle, each almost weightless, but in combination a mass that taxes the ability of the stem to support it. This mass is what concerns us most.

Tefera shows me a test plot in which the stem of each tef plant has been methodically strung through a horizontal rectangle of chicken wire, a sort of trellis that artificially keeps the plant erect. In adjacent plots—without this support—fully half the plants, more in some cases, have simply toppled over, their seed heads soaking in mud. This condition is called lodging and demonstrates a straightforward mechanical problem. As the seed heads mature and gain weight, they overcome the stem's ability to support them and the plant tips its head to the ground. Any gains in potential yield made through good farming are lost to the plant's inherent structural limitations. The chicken-wire plots tell the story. Plants artificially held upright yield 25 percent more than those left to their own devices. But for most farmers, buying wire to artificially support entire fields of tef is not a practical option.

Tefera wades into a test plot and bends to pull a tipped tef plant upright. Fanned in his hand is the seed head, the hundreds of spikelets, thousands of florets arrayed in the airy panicle. Half the seed head, however, is clogged in mud. Virtually all tef grown in Ethiopia suffers lodging, which is to say this simple condition reduces by a fourth the yield of the country's single most important link to food security. Or at least it seems simple, until one regards the specific conditions of tef and of Ethiopia.

Exotic as it seems, tef is simply a grass, *Eragrostis tef*, one species of a large genus of wild grasses, themselves a member of the larger family of Gramineae, grasses, a lineage that places the problem firmly in the family of man. Those who joke that grass is for cows and golf courses have not fully considered the matter. Maize, wheat, and rice, those three grains that provide more than half of human food, are all grasses. So, too, are barley and sorghum. Humans know how to grow grass better than any other kind of plant. The Green Revolution itself was marked by an increasing reliance on grasses. Indeed, if there was a key condition that shaped the Green Revolution, it was lodging, the first hurdle faced by plant breeders as the increased use of fertilizer and development of higher-yielding varieties began, by definition, to add to the weight of the seed head on rice and wheat. Plant breeders came smack up against the limitations of the plants' architecture. Logically they began tinkering there first, reducing stem height, increasing stem thickness, shoring up the little towers charged with supporting the exponentially increasing bulk of the world's population, all of which ought to be good news for tef. So why not repeat a few steps that conquered the same problem in rice and wheat a generation ago?

This notion needs to be considered through the lens of a 32-power microscope, the level of detail that begins to tease out the dimensions of the problem. In his lab, Hailu Tefera has tipped a potted tef plant to splay its seed head across the microscope's field. He peers through the eyepiece while probing, a long needle in one hand, tweezers in the other. He's quiet for a few minutes, concentrating, then announces success and leans back to give me a look. At the tip of the needle, a single spikelet has been peeled from its sheath. Protruding, just barely visible at this magnification, are three reddish nipple-like tips that are the floret's anthers, their male parts. These are the targets, structures smaller than the needle's point.

Tef, like wheat and rice, is self-pollinating, which is to say, each flower contains male and female parts. Each plant head contains as many as 6,000 possible mates for each floret. There is no cross-fertilization; variation comes from mutation. If a plant breeder wishes to tinker with this system, wishes to crossbreed lines by introducing genes from another plant, then the first step is to prevent a plant from fertilizing itself, which is what Tefera is doing under the microscope. He is emasculating individual florets by removing the anthers, those minuscule parts that are barely visible at 32-power magnification.

It is not enough to remove anthers from one spikelet, however, because anthers on adjacent spikelets can get the job done. Not all of the several thousand spikelets are mature and ready to fertilize at the same time, but a wave of maturity spreads up the head, so at a given moment several hundred are, which means that to emasculate a plant one must painstakingly tease out and remove several hundred anthers. Then the cross can be attempted. After a half day's work—if Tefera is skillful and has not damaged too many florets in the process of defusing the male parts—he crossbreeds twenty florets. From these he would expect to get a first generation of ten seeds, all ten of which would cover about as much space as one of the shorter words in this line. These he grows to see what happens, and continues the line for eight generations.

Tefera's breeding experiment is working from a foundation of 320 different germ-plasm lines, the split of the diversity of tef as it exists in Ethiopia. That is to say, after collecting 2,500 samples of tef from around the country, researchers were able to sort them into those 320 types, each with distinct characteristics, the variation in the pool. From this pool Hailu has teased out the most interesting varieties and crossed them to breed 165 lines, each bred one magnified anther at a time. The hope is that within these 165 crosses, some varieties will develop significant resistance to lodging. But only the growing can tell.

Results from other projects have been discouraging. Researchers who have begun analyzing the genetic code of tef have found that, in the language of this work, "tef has a very low level of polymorphism compared to other cereal crops." Not surprisingly, a self-pollinating crop that evolved over a relatively limited range of conditions turns out to exhibit very little variation. Breeders want to exploit the crossing of diverse lines to bring about change, but a very low level of polymorphism means there is little diversity to exploit. Unfortunately, this is only the first hurdle.

As a breeder, Tefera will know if he is on to something only by growing it, and that, as we've learned, is a laborious process involving eight generations of 165 lines and test plots spread around the country to provide a range of environmental conditions. The test plots must be weeded by hand. They must be guarded against pests and, significantly, neighboring farmers. (Researchers found early on that once word got around that the isolated plots might contain improved tef varieties, farmers would steal in and snatch a few handfuls of seed—a backhanded bit of good news in that it signals that any improvements breeders make will probably be quickly adopted by farmers.) The entire process of breeding is expensive as well as laborious, and much of it could be avoided by molecular mapping, by identifying the region of the genome responsible for the traits in question and zeroing in on changes in the region. Researchers would not need to spend so much time confirming variations by growing them in the field. They could read the evidence of change in the genetic map. It now takes ten to fifteen years to develop and release a new variety, but Tefera believes molecular marking could cut that time nearly in half.

The rub is, the research facilities in Ethiopia are not capable of this sort of work. Molecular marking is in this sense inappropriate technology, but in another sense appropriate, in that identifying a molecular marker would allow a poor country like Ethiopia to forgo the expense of field testing. The end result is still a seed, and improved seeds are the most appropriate form of technology.

This was exactly the sort of issue that the partnership concept built into the McKnight program was designed to solve, and so the concept was deployed in Ethiopia. But it faltered. Ethiopian researchers paired up with the Institute for Biotechnology at Texas Tech University, which is why Tefera was in Texas for some added training in molecular-marking techniques. Beyond training, though, postdoctoral students at Texas Tech were to do the molecular mapping.

Tefera says they didn't. Instead, he says, the partnership really generated a revolving door for post-docs, an issue of more than particular interest to this case. Much of the gains that will be made from here on in must be made in "forgotten" crops like tef, while at the same time the basis of the work will be highly technical, such as genetic marker–assisted breeding. These conditions require close cooperation between developed-world science and developing-world needs. In the case of tef, though, asking a post-doc to commit to doing fundamental research on a crop no one had heard of is something akin to asking a budding young American musician to specialize in Outer Mongolian funeral dirges. The work might be interesting, but it doesn't provide much of a platform on which to build a career. An academic institution might not only be willing to put people to work on a problem, it must be able to recruit and retain them.

In this case, and presumably in others, an added dimension emerged—more subtle, but an indication of how academia has evolved to chase dollars. Academic money for general research is scarce, but some institutions have learned how to specialize and chase grants. They have become grant factories. Observers say Texas Tech is just that sort of place, interested in pulling in the funding but lacking the institutional commitment to work long-term on the problem. Some institutions are simply interested in winning grants by telling foundations what they want to hear; some even want the grants to do good science, but even this is not enough. There must also be a genuine commitment to bringing that science to bear on the problems of places like Ethiopia.

In 1998, at the halfway point in the McKnight program, McKnight's oversight committee precipitated a "divorce." Tefera went shopping for a new partner, and eventually signed on with a research group at Cornell. For those running the program, the experience was a hard lesson in the new realities of agricultural research. The situation almost demands a sort of technology of partnership, of allowing people to work together across broad cultural gulfs. No one has yet developed that technology, so muddling through will have to do, at least for now. Meanwhile, although the new arrangement may well clear away some of the institutional problems, deeply imbedded obstacles remain.

For instance, one U.S. scientist who was considering joining the partnership studied the issue at length, then offered a stark assessment to McKnight: "I don't know if this is a crop where you can make scientific progress." Tef is, so to speak, a tough nut to crack, which shouldn't surprise anyone by now. The scene before us— Tefera bent over his microscope—is the proper backdrop for considering a general rule, which is a problem not just for tef but for the whole range of food security issues. The easy problems have been solved. The tef phenomenon is not so much an island as an example for all of us: the low-hanging fruit has been picked. The remaining problems are as complicated as reading whole genomes and the even more complicated dimensions of human relationships.

Along the road to Debre Zeit we stop near a mud hut to visit a farmer cooperating with Debre Zeit's researchers in an experiment in crop rotation. Engda Kelkle has about five acres of land and seven children. Arrayed in a sports coat, as is the habit of Ethiopian men no matter how poor, he walks his fields of tef with considerable pride. They are doing well. Through a translator he gives credit for this fine crop to his own broadmindedness, which resulted in his agreeing to help out with the research. Kelkle adds that tef is 90 percent of his household's food, so the temptation for these small farmers who tend the millions of such plots that frame Ethiopia's plains is to grow tef on tef on tef, which, in turn, raises problems with fertility. It's a form of monocropping, so that the specific demands of tef grown continuously exhaust the soil. Kelkle, like most farmers, has livestock. He feeds the animals when he can, especially tef straw, but their manure is used for household fuel and therefore is not available for the fertilizer that could offset some of the effects of monocropping. Instead, Debre Zeit's researchers have convinced Kelkle to try a rotation of tef with chickpeas, a legume that fixes nitrogen from the air to fertilize the soil. He says it has worked impressively, giving him a bumper crop of tef. He seems a happy man with all this, and it's easy on this gentle green plain to ascribe a certain settled state to this farmer's life—never mind the seven children and the fact that Kelkle's five acres have no prayer of feeding his children's children.

"It is just a miserable life," Tefera tells me. This comment comes as we are leaving the farm, and Tefera seems anxious to depart, just as it was clear to me he didn't want to stop in the first place. We visited a farmer only at my insistence. At first I read this as a status issue, that a scientist with a Ph.D. doesn't associate with farmers in Ethiopia, and maybe that is it, in part, but it is also more complicated. It's not that Tefera doesn't know farms. He grew up on one like this. "To this day, when I go to visit my father, he kills a sheep," he says.

I tell him there must be something working properly in a country that can provide an educational pathway for a kid raised in a mud hut all the way to a Ph.D. in the United Kingdom. He thinks what I've said is naïve, points out that in Ethiopia there is no other pathway, since everybody starts out poor and rural. Most stay that way.

Days later, I've arranged to meet Tefera in Addis Ababa. It's lunchtime, so he picks a place, a cafeteria at the graduate school. Meals here are subsidized by the government, so a lunch of stew and *injera* costs substantially less than a U.S. dollar. Elsewhere in the city, lunch is a dollar, even more, and he cannot afford to eat in those places.

Tefera is the plant breeder primarily responsible for improving the country's most important crop. If he succeeds in cracking the lodging case, he will be his country's Norman Borlaug. His government pays him the equivalent of $200 U.S. a month. It's not that $200 has all that much more buying power in Ethiopia. The subsidized cafeteria meals fit the whole picture. He lives in government housing and estimates that were he to have a life that included such luxuries as his own apartment and maybe a television set, he'd have to earn the equivalent of $1,000 U.S. a month.

Tefera says that most of the young Ethiopians sent abroad for advanced training find higher-paying jobs elsewhere and stay abroad. For example, he worries that an Ethiopian now training in Texas will not return because his stipend as a grad student is more than he can ever hope to make in Ethiopia. Since the Dergue, Ethiopia has opened itself to business and is developing. But in a sense, that has only made matters worse. Business and nongovernmental organizations within the country are attracting people away from core government-sponsored ag research, such as the tef project. Arguably, those people still work for some aspect of improving Ethiopia's lot, but ultimately both the aid organizations and business have different research agendas.

Even more subtle but as draining is the culture's attitude toward the civil service, which is hierarchical. Status, perks, and salaries attach to the administrative jobs, so scientists are drawn away from basic research work to status jobs behind desks. One former plant geneticist, now an administrator, sadly showed me his lab coat hanging in a corner of his office, where it had hung undisturbed since he'd taken his administrative post. In a country where the intellectual infrastructure consists of a handful of trained scientists, the talent pool is quickly siphoned off.

One day Tefera and I drive to downtown Addis Ababa, to a multistory office building among marble banks and hotels and on a boulevard crowded with grazing goats. We walk up marble stairs, talk our way past a secretary to a spacious office with two phones held by Beyene Kebede, head of the Agriculture and Environment Department of the Ethiopian Science and Technology Commission, which makes him the official in charge of all agricultural research in this agricultural nation.

Kebede, gracious and forthcoming, launches straight into a discussion of improvements in research, by which he means a rewriting of the flow charts. This formerly independent agency has been subsumed by some other one. Administrator X now answers to Bureaucrat Y. All well and good, but who will staff this elegant flow chart in the face of brain drain?

"There is a big attrition rate. That is fully accepted," he says.

I try a blunt American reporter–style question: What would be wrong with paying salaries to researchers comparable to what they could make elsewhere? He looks at me a little blankly, as if the idea is too foreign to be considered, and after a bit he tells me such a thing would be impossible. Why? Because it is impossible.

Tefera has explained to me that he came into the system as a young man because that's where the opportunities were. There were openings in tef research, and with those openings came opportunities for more and more training. Opportunity and a certain sense of pride in his people and concern for their well-being. The pride shows through whenever he tells me about tef. Yet he seems poised at the breaking point in this system, all trained and invested in a difficult, long-term, and thankless task. The further he and others probe into the DNA map of the problem, the clearer it becomes that there will be no monumental breakthroughs, only years of breeding and incremental gains in yield. Good, solid gains, but incremental. So finally I ask him if he, too, will become part of the flight to desk jobs or to the high-paying private sector.

"If the opportunity comes," he says, "then surely I will take it. It is a problem, and for the moment, there is no solution for it."

And what then would become of the lodging problem and tef? By way of an answer, he introduces me to a couple of colleagues a few notches below him on the opportunity ladder. Fufa Hundera and Tirunah Kefyalew are both smart young men and tef breeders working on doctorates. And behind them are younger students working on master's degrees and on tef breeding. It is hard to imagine why someone might follow Tefera's path, but as he pointed out, in Ethiopia there is no other path.

The Mengistu years decimated Ethiopia's intellectual infrastructure, but it is being rebuilt. True, it is being drained, but part of that drain is into positions necessary to build even more structure, and part of that drain is being met by an increasing flow into the system. Tef is a stubborn problem emblematic of Ethiopia's larger problems. It is comfortably romantic to think of a single researcher bent over a microscope cracking the case. But it is closer to the truth that the case won't so much be cracked as it will be nudged along. Nudging takes years of institutional commitment. Nine test plots; 160 breeding lines. This is more than the design of research. It is the work, and good work can become an institution in its own right. Properly done, good work, in its own right, can become the glue that ties people together.

Lucy, the 3.2-million-year-old hominid fossil, has led a controversial existence ever since Donald Johanson dug her up in 1974. Science sees her as part of the human legacy, but Ethiopia claimed national rights. Ethiopians don't call her Lucy, a name Johanson's crew took from a Beatles tune popular at the time. In Amharic, she is Dinquinesh, "Thou art wonderful." The Ethiopians have kept her locked inside a piece of the same creaking primitive infrastructure that holds the rest of the country. She is now in the national museum, a somewhat grandiose term for a house the nation inherited at the end of the Italian occupation during World War II. The house looks today like nothing so much as a run-down Depression-era school building.

Lucy is inside, in the second room, resting in a cheap display case behind a hardware store cylinder lock. A typewritten sheet Scotch-taped to the case explains her significance. This represents one perspective on Ethiopia's ability to preserve a legacy. There is another.

My last stop in Ethiopia, another multistory office building, the Institute for Biodiversity, had goats grazing in the yard. In the lobby hung a portrait of the Russian geneticist N. I. Vavilov on his visit to Ethiopia in the 1920s. To a plant breeder this is the equivalent of a consecration of holy ground by the Pope himself, a visit that formally marked Ethiopia as having been one of twelve Vavilov centers in the world, centers of biodiversity. In the 1970s the world became greatly concerned about genetic erosion and preservation of the legacy that was the basis of agriculture, so an international movement led to aid being granted to Ethiopia, especially from the German government, to establish a germ-plasm bank. This is it.

The institute's director, Abebe Demissie, a plant breeder, met us in his office, but didn't so much want to air his flow charts as he wanted to show me his seeds. We toured a series of seed labs and storage facilities as capable and up-to-date as any—56,000 accessions, 101 species of everything from cereals to medicinals to coffee, all cataloged and lined in neat rows in freezers. If there are solutions to the problems of tef and to those of Ethiopia's other crops, they will be teased out of this diverse genetic legacy. A lot of havoc, destruction, and chaos may reign on the streets outside. Its anthropological legacy may rest in a rickety case, but the country's genetic legacy lies stored in the bank, helping to secure a future for Lucy's line.

# How Things Fall Apart

## When Politics Pushes People Against Nature's Limits

### [Zimbabwe]

The essential elements of rural subsistence life dictate a kind of geometric order. Like many such houses around the world, Ndlovu Mancube's hut in Zimbabwe is round, the walls arced on a central point that is a fire pit. One guesses this architecture flows from the logic of even distribution of heat; the design of huts evolved from the design of campfires. Geometry places the cooking pit at the center of life, the position the struggle for food occupies here in a village in the Matobo region of Zimbabwe and in most of the poor households in the world.

As it places food at the center, the fire pit sends people to the perimeter, farthest from the smoke. So we visitors sit along one wall on benches and stools offered us; Mancube and a semicircle of children, a representative sample of her eight and the three grandkids she raises, arrayed along the arc opposite the visitors. Her husband is absent, as are most husbands from the surrounding similar circular huts, most off on business ventures or whatever. Explanations of these absences are usually vague. Mancube's husband works at the nearby national park, the same park that brings long, vociferous complaints from most of the villagers one meets in Matobo: it fences out their cattle. Not so many complaints here in Mancube's hut, though, in that the park is in part responsible for a certain measure of their prosperity. The walls, neatly stuccoed and painted, come complete with heart-shaped masonry shelving that makes up the kitchen alcove. There are stacked a neat array of bowls for *sadza*, the sorghum or maize porridge that is the staple of Zimbabwean life, rural and otherwise, subsistence and otherwise. The concrete floor is troweled glass-smooth and painted.

Part of their well-being comes from Mancube's husband's income from the park and part directly from the tourists the park draws. Some of the children of the house have become expert carvers and spend their days whittling ornate wooden bowls, many of them formed from finely wrought figures of wild animals. They're made from the acacia trees cut from the surrounding hills. Each bowl brings the equivalent of a couple of U.S. dollars.

The family's support also rests on Mancube's considerable skill as a farmer. In Matobo, as in much of Zimbabwe, men do not farm; they care for cattle. These days, with the cattle gone, the men do other things, but the women still farm. Evidence of Mancube's skill is clear on her land. We can see, for instance, the manure piles on her fields, the hectare or so of land around the hut. These piles were what drew the visitors from the nearby road, because manure is a prime indicator of a relatively prosperous farmer in this region. Most farmers have no livestock for manure, or if they do, they burn the manure for fuel, because they can't afford charcoal. Mancube also has a small orchard, a clutch of carefully tended fruit trees—peach and guava—springing from what one would think, especially now in the dry season, a desert. The trees grow in hard-beaten sand. Spent soil and aridity make up the primary parameters of farming here. The Mancube family is relatively prosperous, although only when seen against the general backdrop of the poor people and poor soil all around.

The McKnight project in Zimbabwe deals mostly with sorghum, another way of saying it deals in subsistence agriculture, which is another way of saying it deals in places like Matobo. All these things go together. Like round houses, the design of the Zimbabwean land-scape is dictated by agriculture, by agriculture and, of course, by colonialism. In Zimbabwe, sorghum draws the line, sets up the dynamic that can make the difference between a miserable existence and a tolerable life. At 390,580 square kilometers, the nation is slightly larger than all of the British Isles. Landlocked, it sits just south of the equator, a largely flat, dry plain that holds about 11 million people. Mostly pleasantly temperate, it has an easy climate, an assessment that can be reinforced with a visit in the austral spring, when the jacaranda are in bloom. Zimbabwe is a grassland and largely suited to agriculture, but not uniformly so, which is where colonialism figures in the geometry.

Mancube is not Shona, Zimbabwe's ethnic majority, but rather Ndebele, the country's other major ethnic group. Her home lies in the south near the sprawling modern city of Bulawayo. A few minutes' drive from her house, within Matobo National Park, the reserve that so vexes the locals, lies the grave of Cecil Rhodes, the British colonist who made his fortune in South Africa's diamond fields, then ventured north to begin laying claim to more land, planning a broad corridor of Victorian empire stretching north to Cairo. He cut a deal with the Ndebele chief Lobengula in 1888 at Bulawayo, which had recently become the Ndebele capital. In exchange for a riverboat, 1,000 rifles, 10,000 rounds of ammunition, and 100 pounds sterling a month, the Ndebele king gave Rhodes a foothold in the territory that would become Rhodesia.

Rhodes imagined he was gaining mineral wealth, but what he found as he ventured north of Bulawayo was farmland, farmland occupied by the Ndebele and Shona, who together still make up more than 95 percent of Zimbabwe's population. The country's gifts are not uniformly distributed. A broad central region gets the bulk of rainfall, while the fringe, the geographic and economic perimeter of the nation, is near desert-dry. The most productive lands went to the colonists who followed Rhodes. Today these lands raise cash crops on large-scale commercial farms: maize, cotton, coffee, tea, tobacco, wine grapes, livestock, citrus fruits, and vegetables.

Eighty percent of the country's population still depends on agriculture. Most of the people who depend directly on agriculture are subsistence farmers, who wound up with the most marginal lands, while the cash-farming colonists got the best. When Rhodesia became Zimbabwe at independence in 1980, the issue of land distribution was placed center stage. Robert Mugabe, still Zimbabwe's political leader, came to power promising to move 162,000 blacks onto white-owned lands, but negotiated with the British, as a condition of independence, a clause that said whites would not be forcibly removed from their lands. There has been a mass exodus of whites from the nation, so that they now constitute only about 1 percent of the population, about a third of what they did at independence. Whites nonetheless own a third of the country's arable land and produce about 42 percent of its export income. In a country strapped for export income, that last number does more than the rest to cement the present system in place.

Doreen Vhevha has the sort of organized mind that can produce a flow of conversation that works like hypertext. She does not offer arguments so much as she constructs them, methodically, looking off into the middle distance as she retrieves the relevant bits and pieces. A seemingly random and inconsequential seven-digit number she hears once on one day comes back on demand a day later, as do quotes, names, and concepts. She's a social scientist by training, angling toward a master's degree in economics. She's also a young mother and, like many urban Zimbabweans, only a generation removed from rural Shona life and some of its customs, such as the responsibility to clean her mother-in-law's house. Unlike many scientists who come to agricultural issues as scientists first, fascinated with the nuts and bolts, Doreen is unabashedly altruistic. She is first and foremost interested in helping the poor of her nation, the farmers.

"We've got to help these guys," she says. "The only way you can actually develop Zimbabwe is to increase the income of the majority of the people, and the majority of the people are farmers."

One of the charts Vhevha lays out details the taxonomy of the landscape according to its natural gifts of climate and geology. Zimbabwe is divided into five zones, roughly defined by decreasing annual rainfall. Zones 1 through 3 are best for farming, the colonists' land, devoted to agribusiness. Zone 4 is semiarid, and 5 is suitable for not much more than extensive ranching. The McKnight project works solely in zones 4 and 5, simply because that's where the poor live. Aridity draws a ragged border around subsistence.

McKnight's decision greatly limits the range of possibilities, but unlike subsistence agriculture elsewhere, it places the work in areas largely lacking agricultural precedent. Unlike the maize growing in Mexico's Sierra Norte de Puebla or the tef growing outside mud huts in Ethiopia, Mancube's guava and peach trees have been growing in sand for only the century since colonialism set them there. The Zimbabweans running the project accepted the colonial resettlement and have elected to work with conditions as they still are. "What is most important in this country is that we improve the lives of the local people, and we improve the lives of local people by working with what they have," says Vhevha.

More than a pleasant truism of the sort that has become boilerplate on grant applications, this decision marks a real fork in the road. Unfortunately, it also marks one of those turnings that in Greek tragedy would signal a "fatal flaw," a sign that this story, at least in one sense, is going to end badly.

A good botanist, taken blindfolded to any corner of the planet, should then be able to read the political history of the place from his first look at its flora. In Zimbabwe, the struggle, on marginal lands, between two plants—sorghum and maize—tells something of the history of colonialism. Sorghum is the African cereal, maize the settler, brought in by Rhodes's compatriots and promulgated throughout the continent by the miracles of the Green Revolution. The resulting tension manifests itself even in the national dish, the ubiquitous *sadza*, once made of sorghum but now most often consisting of white maize. The fact of the matter is, Zimbabweans have cultivated a preference for maize, as demonstrated by the cold economic fact that kilo for kilo, maize costs more.

To the average subsistence farmer this means an even more tenuous hold on food security. Farmers want to grow maize on land that is better for growing sorghum. The soil is too poor and rainfall too meager to support maize consistently, but every once in a while somebody gets lucky: hard rains come, the timing is right, and somebody out there on the edge brings in a bonanza of maize. Maize's high price and high yield make the gold-strike metaphor appropriate from the vantage point of a farmer who lives in a round hut. Maybe this good fortune strikes in the next village, but still, everyone hears about it and a maize crop becomes the grail. Planting it is a gamble, but a gamble Zimbabwe's subsistence farmers willingly take year after year, always looking for a new combination of seed and fertilizer that might tip the balance their way. (Zimbabwean subsistence farmers will spend money on fertilizer for maize, but not for sorghum. In interviews with farmers, researchers often asked them why. The farmers answered that they didn't know there was such a thing as a fertilizer that worked on sorghum. The practice of fertilizing was introduced with maize.)

"They want to grow maize, and every year maize fails and they starve," says Vhevha. The fissure between maize and sorghum, then, was where the researchers saw opportunity.

You may recall that sorghum contains, to varying degrees, compounds called tannins. The higher the tannin content, the browner and more bitter the sorghum—and the less nutritious, because tannins bind to proteins, making them unavailable for digestion. No

45

species knows this better than quelea birds, the sparrow-sized seed eaters that constitute a pestilence of biblical proportions throughout the vast strip of East Africa where sorghum is grown. The birds dislike brown sorghum. Elsewhere in Africa, as I already mentioned, some people have exploited this trait by growing the high-tannin sorghum that birds disdain and processing it with wood ash to remove the tannin and make it more palatable to and nutritious for humans. In Zimbabwe, however, this does not happen. Researchers know this practice is used elsewhere but say Zimbabweans simply won't eat *sadza* that tastes like ash. Instead, they grow lighter varieties of sorghum for porridge and deal with the bird problem by sending children to the fields to chase away the birds—yet another reason why sorghum has lost favor.

"Who will stay out in the fields now instead of making these wooden things?" Mancube asks, pointing to the wooden bowls her children have carved to sell to tourists. We were to hear her explanation again and again in the villages of the Matobo district. Sorghum is no longer grown because no one has time to chase away the birds.

More is at stake in the relationship between these birds and these people than simply whether the humans will eat porridge of maize or of sorghum. The real issue is the relentless pressure on having enough to eat today. It forces families to pull kids—girls especially—out of school to chase birds, to carve bowls . . . to make more children. So whatever is grown, there are few educated children who might in time improve the food supply.

Vhevha has already told me there is a saying in these villages—a girl should go to school just long enough to learn to write her first love letter.

The decision to work with what the poor farmers of Zimbabwe already have entailed first learning exactly what they did have through intensive focus group–style interviews in a number of villages that represented a cross section of both the Shona and Ndebele regions.

Researchers used speakers of both languages as interviewers, but took pains to dress like and otherwise to blend in with the village's women. They attempted as much as possible to adopt their accents, inflections, and habits. They also interviewed women and men separately. The men participate in some farming chores, but traditionally have stuck to handling livestock. The women were more forthcoming on their own, and also had specialized knowledge the men didn't.

Researchers found that, although the farmers had no actual word for tannin, they were fully conscious of a wide array of characteristics related to tannin. In addition to their strong preference for white sorghum and their knowledge that birds like the same kind of sorghum people do, the villagers knew that high-tannin sorghums work well for traditional beer. In addition to *sadza*, a staple of village life is a crudely fermented sorghum porridge named according to the length of its fermenting time. This is not as licentious as it sounds, in that the fermentation process enhances some of the nutritional qualities of the grain. A beer fermented just once is called *jwanday*, and anybody can drink it, but a seven-day variety is an adult matter.

The villagers also had a complex knowledge of varieties of sorghum according to other qualities, preferring, for instance, early-maturing varieties, which makes sense in a region with a long dry season often marked by famine. A couple of names for early-maturing varieties encode this hard-edged significance: *mukadzidzoka* means "wife come back" and *mukadziwaenda* means "wife don't go." Wives leave houses where there is no food.

The villagers' preference for maize was clearly based on a rational sort of gamble, and they understood that market forces had brought them to their precarious position. Although they are mostly losing the gamble, it was evident that simply telling them to grow sorghum wouldn't sway them. They needed to hear it from the market.

In the face of these realizations, the researchers developed a multidisciplinary, multifront attack focusing on tannin. Tannin is not a poison but rather a fence. It keeps the birds out, which is its service, but it also keeps the people out. Research focused on ways to make that fence more pervious to people, a straight-on run at the problem. The point person for this strategy was Trust Beta, a lecturer in food science and nutrition at the Institute of Food Nutrition and Family Sciences at the University of Zimbabwe. Tannins have such a dramatic effect on nutrition that rats fed a diet of high-tannin sorghum will lose weight, so completely does tannin bind proteins. But Beta points out that this case is extreme. She has analyzed seventeen varieties of sorghum grown in Zimbabwe and found that the real-world range of tannin concentrations rises nowhere near the levels fed to the rats. That is, what passes for high tannin with Zimbabwean farmers and Zimbabwean birds is still relatively nutritious served straight up. Indeed, bitterness has conditioned farmers and birds to avoid brown sorghum, but there are other problems associated with tannin.

For example, the high-tannin varieties tend to be soft and to break up during hulling, so that some of the grain is lost with the hull. Beta is working on preconditioning methods that cut these losses. She is also working to pursue avenues aimed at making sorghum more palatable to industry, such as working with the brewing industry—which now uses formaldehyde to combat tannins—to find other processes less objectionable to consumers. She is also seeking ways to process starch from sorghum. She rationalizes that anything that will raise the price of sorghum will cause farmers to grow more of it, boosting their food security in the bargain. Chickens are a key element of this strategy.

American political tradition preserves the phrase "a chicken in every pot," but there isn't a way of preserving the resonance the language had in the time it arose. In the early part of the twentieth century, especially in an American South still rooted in subsistence agriculture, a chicken symbolized economic well-being. In Zimbabwe, Vhevha says, chickens maintain that status. In the center of the kitchen of a spotless hut near Matobo, with a slick-painted concrete floor and a meticulous assortment of plates along a kitchen shelf, a neat wooden box rested, altarlike, and held a sitting hen.

Chickens make up the centerpieces of feasts and are prominent players on many special occasions.

An old American blues song advises that "a chicken ain't nothin' but a bird." Maybe, but unlike quelea birds, chickens will eat sorghum—a fact farmers don't seem to know. They think that if queleas avoid brown sorghum, it must also be bad for chickens. The substantial Zimbabwean chicken-feed industry, which at present runs on maize, doesn't know it either, but researchers have shown that chickens do well on an exclusive diet of high-tannin sorghum and on diets of mixed sorghum and maize, just as well as on straight maize. This bit of information opens up a couple of possibilities: farmers could grow high-tannin sorghum and feed it directly to their own chickens, and the feed industry might also be persuaded to blend the cheaper sorghum with maize, thus increasing demand for sorghum. These possibilities, though, open a broad line of questioning that leads to the work of John Dube.

Among the young researchers and students in the project, Dube is a gray eminence, but everyone calls him J.S. He's an easygoing man who speaks as if everything he says has been carefully considered, and who has an engaging and ironic sense of humor. He's a researcher at the government-run Matopos Research Station, a sprawl of land just south of Bulawayo that looks like a big ranch in the American Southwest. If the buildings appear a bit colonial, there's a reason. The land was once the estate of Cecil John Rhodes, who willed it to the government as an experimental farm. It became an agricultural school in 1921 and an experimental station in 1934. Its arid location sets a research agenda based in rangeland and livestock issues. In addition to the government, the station has nineteen partners, including the International Crops Research Institute for the Semi-Arid Tropics (better known as ICRISAT in the acronym-choked field of international agricultural research) and the International Livestock Research Institute.

Officially, J. S. Dube is listed as the principal research officer in chemistry in all of this, but he directs his chemistry at trees. One can get a significant clue as to why this might be a fruitful area of investigation by taking a quick drive through a game preserve just to the north of the research station. Here a Discovery Channel–type array of African fauna grazes and browses on a dry-season savannah. J.S. noses our rented Mazda off the dirt road. We had been taking turns driving all day as he guided me around the Matobo area, because neither of us was very good at it: I drove on the wrong side of the road and he couldn't get the hang of the manual transmission. Still, he thought it worth the extra grief to find giraffes, and eventually we saw the gangly, ungainly beasts working their way through a line of acacia trees.

I told J.S. the giraffes must have been his first clue. If evolution went to all the trouble of designing such an unlikely structure, just to reach the upper limbs of acacia trees, there must be something up there worth having. And there is.

Acacias are legumes. Trees, true enough, but pod-producing members of the pea family, like beans. Legumes, as we have seen, are the basis of multicropping. They are nitrogen fixers, pulling free nitrogen from the air and fixing it in their leaves, where it is then available for animal nutrition or for working into the soil in order to fertilize other plants. The American agronomist Wes Jackson says agriculture ought to copy the design of the American tallgrass prairie, because the prairie can "sponsor its own fertility." Legumes are the primary sponsor.

If acacia trees are nothing so much as leggy legumes, and if they are the dominant vegetation in a region where the understory is more or less stripped to sand, then they have a part to play, a considerably larger part than simple animal nutrition. Just to thicken the plot, though, it helps to know that the acacias contain tannin in varying degrees.

The obvious use for the trees is as forage. Dube and several other researchers are looking in detail at six species of the genus *Acacia*, both natives and exotics, all existing in the regions under study. The tannins in the trees present the same set of problems as those in sorghum. The levels of tannin are high enough in some of the foliage to make the trees unpalatable to animals such as goats and hogs. The researchers have found that simply by mixing leaves from various trees, they can come up with a palatable product. They're now at the point where they can hand out a simple prescription to farmers.

This work is directly relevant to food security, because it provides people with protein from meat, but also makes livestock a part of the complete farming system by giving farmers a badly needed supply of manure for cereal crops. A direct tie to soil fertility is also evident. The foliage from the trees is rich in nitrogen, and the mulch is a fertilizer. As this mulch decays, it builds organic content in the soil, making it more like soil instead of sand. It leaves voids that will hold water. It turns out that tannins play a key role in mediating the flow of these nutrients as well, a complicated relationship and part of the research. Just as in animal nutrition, tannins restrict the ability of soil to "digest" nitrogen. A goal of this line of research is to develop a set of methods for mulching attentive to all these complications. Just how far this element of the work might go becomes most striking in a screened-off experimental plot on the University of Zimbabwe's campus.

Rosemary Chanyowedza has an easy laugh that rolls around the "screenhouse" that holds her work, the basis of her doctoral dissertation. She shows us a series of weedy test plots without weeds; the difference can be explained by allelopathy, the ability of some plants to secrete chemicals toxic to others. Humans are latecomers to the herbicide business, entering a field already developed by plants themselves. One way in which plants protect themselves is with poisons. Basically, Chanyowedza has treated some of her plots with a leachate made of a mulch from acacia trees. Polyphenols from the acacias—she's not sure which ones—suppress weeds. That is to say, Chanyowedza's work indicates that acacia trees contain a natural herbicide that kills the weeds that attack cereals but allows the cereals to grow, a sort of magic bullet. She explains this significance not in yield gains but in children. In subsistence agriculture, children chase weeds as well as birds. She thinks her work can free some children from field duty.

Chanyowedza mentions that the allelopathic effect of the mulch seems to extend to a plant that can't be weeded, to *Striga*. The mention of this weed pulls me back to a conversation I had at Purdue University with Gebisa Ejeta, an Ethiopian who left his country after the political upheavals that began in 1974 with the overthrow of Haile Selassie. An agronomist at Purdue, he is credited with developing the first *Striga*-resistant strain of sorghum. To explain the problem, Ejeta handed me an article from the journal *Science* that began like this:

> Civil war, genocide, corruption and political incompetence have conspired to keep entire regions of Africa on the brink of famine, earning it the sad reputation as the hungry continent. But even if these dreadful socioeconomic problems were alleviated, African agriculture would still suffer from a host of more traditional problems—insects, birds and plant disease. Indeed, one of the greatest sources of crop losses in Africa is not war or corruption, but three species of the parasitic plant *Striga*.

Crop losses in Africa range from 15 to 40 percent, all due to this single weed, known as witchweed. In the worst areas, it cuts yields by two-thirds, not just of sorghum, its primary target, but of a range of other cereals like barley, millet, and even Ethiopia's tef. One species even attacks cowpeas.

*Striga* is a parasite with a highly evolved strategy of attack. Each plant typically produces about 100,000 seeds, which can lie dormant and viable in the soil for twenty years waiting for a host. The seed chemically senses the host, grows, and attaches a rootlike structure to the host plant to rob it of nutrients. Much of the damage is done even before the host emerges from the soil, meaning farmers have little hope of controlling the damage by hand-weeding.

Ejeta's *Striga*-resistant sorghum is one tool in fighting the weed— an effective one, but the war will need more weapons. It is nonetheless the sort of advance that in a more reasonable world would qualify Ejeta as an international hero.

Facilities to determine which compound from the acacias is suppressing the weeds, or whether it can be synthesized, do not exist in Zimbabwe. Rosemary Chanyowedza hopes to go abroad to finish her research. The McKnight grant was meant to help finance that training and the rest of her research, but the future is less certain now.


We're in the conference room at the University of Zimbabwe with J. S. Dube and a circle of grad students, all earnest young men who talk about livestock. In that it is mostly women researchers who have been talking about plants, food handling, and social issues, one can't help but notice that the traditional gender roles of village life are preserved even in division of research. Just as evident throughout the conversation is that these men have a genuine altruistic commitment to their work.

I try a brain-drain question on them, one particularly appropriate to Zimbabwe, which does not lose its trained young people to other countries so much as it does to the commercial sector within its own. These young scientists are conducting highly specialized research for marginal farmers. Once the educational system has invested in their various agricultural degrees, what guarantees that they won't transfer their training to a high-paying job in industry?

They tell me that while the work is highly specialized, it does tie in rather quickly to the commercial sector, the potential of sorghum for commercial chicken feed being one instance. What such connections present to them, though, is not so much the possibility for their own personal upward mobility as for the mobility of an entire class.

Always in the background of these discussions, the question of land redistribution promised at independence and still a national goal crops up. These grad students see their work as part of that process, as a role they can play in helping farmers bootstrap their way to larger, more commercial farms. It strikes my American sensibilities as refreshingly odd that my question simply does not compute. The only way those students could make sense of it was to translate it into a question of making life better for a whole sector of desperately poor farmers. This notion, then, formed the backdrop for the more unsettling discussion that was to follow.

Tracing the thread of this discussion begins with recalling J.S.'s rumination on browse for goats. Follow it now to a report by Charles Chakoma, who works with browse as well, tackling the tannin issue. Specifically, he focuses on a couple of synthetic compounds, including one in particular developed by an Australian firm. When mixed with foliage from the acacia, these compounds bind with the tannins, preventing them from blocking digestion of proteins. The researcher said he was particularly encouraged in that this compound, the Australian firm assured him, could be cheap enough to be affordable and cost-effective for subsistence farmers.

We must consider all this in light of giraffes, tannin fences, and missing men. Throughout the forage and mulch discussions, the researchers punctuated their arguments by adding that these additional uses of trees would go a long way toward teaching subsistence farmers to value trees, ultimately leading to their conservation. Which causes one to ask, What happened to their valuing of the grass that used to grow beneath the trees?

The farmers around Bulawayo complain about Matobo National Park because it still has grass. Its fence lines form the same stark contrast one finds in semiarid areas around the world—including the United States—between the boundaries of overgrazed livestock lands and those wildlife reserves. The acacias of the region once presided over a rich savannah which raised the abundant wildlife for which this region of Africa is famous. Lightly grazed, this arid landscape could support some cattle, within limits. But the history of Zimbabwe foreclosed on that option by moving a large number of people onto a landscape too arid to support them. The result was widespread overgrazing. Now the grass is gone, as are most of the cattle, and the men who looked after the cattle are off doing other things.

Mrs. Mancube, a middle-aged farmer, reported through a translator: "When I got married, there was water all over the plain and just a few people. Now you can't get water."

She said that after the streams dried up, "some people were disrespectful and dug up the springs. It was sacrilegious." As a consequence, the water went away for good. Science is not altogether clear whether sacrilege does indeed dry up wells, but there's plenty of evidence that overgrazing does.

I ask J.S. if making more forage available to goats was really an admission of failure to manage the grass and a way of guaranteeing that goats and trees will suffer the same fate as grass and cattle. That is, having exhausted one resource, were the researchers and the Australian chemical company simply engineering a way to exhaust another? He answered that grass is not the issue with goats, that they didn't overgraze, because they prefer eating shrubs to grass. The people, however, prefer cattle to goats, and still the cattle are gone.

To say the farmers of the district occupy arid land makes clear that they are pressed against some hard limits, and those limits are hard for a reason. The trees are tall and the giraffe, an odd beast, is evidence that the system makes eating them the exception, not the rule. In a sense, tannin serves the same sort of function as the height of the trees does in preserving the forage from all but the exceptions. Tannin acts as a sort of limitation. Sorghum survives the evolutionary arms race with quelea birds because it has tannin as a defense. So what will survive the evolutionary race with humans who introduce chemicals from Australia to push back those limits for them and for their goats?

On this discordant note, our discussion in the conference room ended, not so much because we had exhausted the topic, but because a relic of British colonialism was to be observed: it was teatime. One of the students reminded me of the irony that tea gets much of its taste from tannin. And so we humans sat around a conference table and sipped of the bitterness that puts most animals off; but we cross the line so readily that we relish the bitterness and celebrate it with ceremony.

Outside, the clean, modern brick buildings of the university's campus soak up some soft spring sunlight filtered through the blooming jacarandas. An air of peace is broken only by the noises of men trenching fiber-optic cable from building to building, wiring them to the Internet. It's not until one realizes what noise is missing that one understands that the peace isn't really peace. There are no students. The university is at the moment a big research campus of grad students and professors. The undergrads have been on strike for four months. Armed guards stand at the gates. Fiber-optic cables notwithstanding, the plain old-fashioned telephone lines don't work. Perfectly good lines link the campus, but at the beginning of the strike they broke and haven't been fixed. It's the sort of circumstance that characterizes much of Zimbabwe, a standout in Africa for having decent roads, schools, clean city streets, an infrastructure of sorts, but also with a regime widely regarded as ossified and corrupt presiding over the country's collapse. The currency is crashing. Rings of shanty towns and tin shacks grow around the cities. There is grumbling and even open resistance to the Mugabe administration, which has turned its back on internal problems and instead spent its millions on a military adventure in the Congo that in 1998 was threatening to push much of Africa into war. Zimbabwe has fiber-optic dreams, but a reality of broken wires.

I walk across campus, then link up again with Doreen Vhevha and am struck once more by her good humor and cheer in the face of the general disarray and even specific disappointments. She has already reminded me that my job as a journalist here is to write an obituary. A few months earlier, the oversight committee for the McKnight program had met in San Diego to re-evaluate all nine of the programs at the midpoint of the six-year project. Eight were renewed for another three years: Zimbabwe was cut.

Robert Goodman, the chair of the oversight committee, says circumstances conspired to cause the cut, but a lack of faith in the Zimbabweans' ability was clearly not a factor. One element in the decision was that the promise of the other programs to make some real gains dictated the logic of allocating more money to them. Once that had been accomplished, there was simply nothing left for Zimbabwe, whose program was not showing as much promise as the others because of its scatter-gun approach to a complex set of issues related to tannin. It lacked focus, says Goodman.

My own interpretation is that the lack of focus comes from the context—marginal lands. Successful farming entails working with nature, but in Matobo, a place nature has already set as off limits, no precedent or natural flow of forces dictates a focused strategy. This leaves only a course of desperation, a clutching at straws. The underlying issue remains land redistribution, and science is not terribly good at solving problems that politics refuses to solve.

Whatever the reasons, the committee's action a world away dashed some hopes in Zimbabwe. It has thrown Rosemary Chanyowedza's work in jeopardy, as it has most of the other research projects. Some of them will go on, finding support from other sources; some will go on of their own momentum. Some of what has been learned will still be put to use.

So what has been learned and what irrevocable changes have been made in the three years of work? The group was unanimous on two points: the terms of the McKnight grant forced interdisciplinary collaborative research, and the process of carrying out that directive had caused lights to go on in heads all over campus. People started thinking outside of the usual boxes, started asking questions of colleagues in different fields, the result being a quantum leap in their ability to deal with problems. The young researchers agreed that there must be a tangible institutional change in the way Zimbabwe does its research.

Waiting for the taxi that would take me to the airport, I wanted to try one more run at the basic question with Vhevha. Wasn't the research program a Band-Aid that ignored the fundamental problem of too many people on marginal, arid lands that cannot accommodate them, of violating the system's limits? The project was seeking a technical fix but ignoring the fundamental social problem.

"What we are hoping is that these people will be resettled on better lands," she said.

But how can there be enough lands when there are ten and thirteen kids in each hut?

"The population problem will be partially solved by AIDS," she said, as if reciting a mathematical formula.

Zimbabwe has an AIDS infection rate approaching 25 percent, among the worst in Africa. Indeed, if population is a problem, then AIDS is, in a ghastly sense, something of a solution. Zimbabwe would have a typical developing-world population growth rate were it not for the disease; AIDS has trimmed it close to Zero Population Growth, or ZPG, to use the environmentalists' acronym. Some who study the epidemic are beginning to draw parallels to the Black Plague of Europe, which killed a third of the continent's population in three years in the fourteenth century. Economists have pointed out that the result was effective land redistribution, so that survivors were much better off financially than before the plague. But a country's overall economic well-being is not wholly dependent on land per capita. The villages of subsistence farmers provide witness to another story.

"People go to the communal areas [villages] to die," says Vhevha. "My grandmother doesn't know a word of English. She stays in the communal areas. She says that before now the children would leave and never come back. Now the children are coming home. They are coming home to die."

In the villages, the children leaving and the young adults dying have meant there is no one to chase the birds. Yet the human impact on the country as a whole has an even more profound social dynamic, one that can be read in a decline in the growth of the GNP. *The New York Times* quotes a Zimbabwean personnel manager as saying it has become his practice to hire three people for each semiskilled job opening, because experience has taught him that two will die during training.

AIDS is a class-conscious plague in Africa. The young who leave the villages to become educated are mobile, and the disease loves mobility. To quote the *Times*: "A well-known 1987 study in Rwanda showed that a pregnant woman had a 9 percent chance of infection if her husband was a farmer, a 22 percent chance if he was a soldier, a 32 percent chance if he was a white-collar worker and a 38 percent chance if he was a government official."

The discussion of food security in Zimbabwe as well as in the rest of Africa is underscored with terms like "bootstrapping." No matter what the specific problems of this grain and that vegetable, a big part of solving them is the task of assembling a native, educated force of young people, the very group AIDS seems to target. The empty university campus at Harare is probably a telling symbol for a country whose brightest young people will die if the disease continues on its present course.

# To Work in Peace

## Visionaries in Violent Times

### [Uganda]

By the time he showed up for his first day of a work at the Namulonge Research Station just outside Kampala, Uganda, Robert Mwanga, a plant breeder newly issued from grad school, had already known troubles. He had been an undergraduate at Makerere University, Uganda's principal university, during the last years of the reign of Idi Amin Dada, when the dictator's army would routinely roll into a village and kill by the hundreds. One day soldiers came on campus and rounded up students, herded them to the third floor of a building, and presented them with a choice of being shot or jumping. A lot of legs were broken, an effective control on student marches.

With graduate work came a chance for Mwanga to study abroad, and he took it, traveling to the Philippines in 1983 only to find himself in a front-row seat for the upheaval and fall of the regime of Ferdinand Marcos. When he came home in 1986 to take a job breeding sweet potatoes at Namulonge, the main government-sponsored research station, his friends asked him, Why are you going there? No one is left.

Mwanga tells me this story in a cavernous, stark, high-ceilinged office at Namulonge, which has no telephone. Our conversation echoes off the concrete walls. The station's buildings are an artifact of colonialism, a former outpost of the British, and they look as if they have not seen significant change since. The grounds are well kept, though, and there is a golf course of sorts, where duffers from Kampala, just a few old folks who acquired a taste for the sport during colonialism, come to play.

Mwanga has no time for golf. He is a dead-ahead kind of guy, slight-statured, earnest, precise, measured, and as efficient as accounting software. He seems almost detached as he tells his story, but once in a while he spreads his long fingers wide and his eyes grow large to punctuate a phrase.

There was indeed no one at Namulonge when Mwanga arrived to begin work. "Most of the staff had run away," he says. "At the time they left, they were leaving hell."

Ugandans and Tanzanians drove Amin into exile in 1979, but in some ways this made matters worse, ushering in a period of warlordism that didn't even begin to settle until 1986, when the current President, Yoweri Museveni, consolidated power. When Amin left, his soldiers remained, becoming, if anything, even more lawless in the vacuum. Whole villages disappeared. A large contingent of soldiers was garrisoned just outside the research station's gate, where they had erected what Mwanga called "structures" whose sole purpose seemed to be to hold the bodies of the executed. "By the time I came," he says, "I just found skeletons."

Skeletons—but the station had no staff, no water. There was a tractor that didn't run and some gutted buildings. Mwanga moved into one of them and did what he had been trained to do. He began collecting germ plasm, samples of 400 separate types of sweet potatoes, both local landraces, or locally developed varieties, commonly grown by farmers in Uganda, and varieties from elsewhere. There was work to be done.

In seven countries in the region—Uganda, Burundi, Congo, Ethiopia, Kenya, Rwanda, and Tanzania—sweet potatoes are one of the top three food crops. Uganda is generally the largest producer in Africa and fourth in the world. Only China, Indonesia, and Vietnam produce more.

Introduced by colonists and missionaries only a century ago, this plant native to the Andes does much to support Uganda's 17 million, along with bananas and cassava. The nation's lush, tropical rolling hills are tiled border to border with two- and four-acre subsistence farms, and virtually every one of them grows at least these three crops. Kampala's burgeoning urban population is fed by a constant stream of trucks that carom the potholed highways, freighted high with loads of white and amber sweet potatoes.

Yet Uganda, among the best of these African nations at raising this crop, still gets yields of only about 4 tons per hectare, while yields of 75 tons have been recorded elsewhere in the world. Some of the poor yield can be blamed on the primitive state of everything in the region, agriculture included. More directly, though, a complex of viruses along with a persistent sweet-potato weevil is putting such pressure on production as to cause near-total losses. Mwanga says that even in field trials using the best available practices, he's seen losses of more than 60 percent to the virus alone. In real-world conditions, 90 percent losses are not unusual, which means a subsistence farmer can lose 90 percent of her family's food.

Mwanga was not as alone as the abandoned research station would indicate. Sweet potatoes are an international crop with a legacy of research. Their various problems have been tackled by land-grant schools in the southeastern United States, where they are a major crop. The International Center for Potatoes, known by its Spanish acronym CIP in Lima, Peru, is in the Andes, the center of origin for both potatoes and sweet potatoes, and it coordinates a worldwide network of knowledge, knowledge distilled to and codified in germ plasm. Both weevils and viruses have long plagued the world's sweet potatoes, and breeders have successfully tackled these problems. In addition to collecting local landraces, Mwanga tapped the international system.

"When we started, the people from CIP said, 'You don't have

to do the breeding,'" he says. CIP already had bred and released virus-resistant strains.

To a plant breeder, especially a breeder working with the idiosyncrasies of the Ugandan system, this is not so much a solution as a beginning. East Africans happen to like and eat what the rest of the world would regard as pretty unpalatable sweet potatoes. They are white and cook down into a starchy, relatively tasteless paste. Ugandans call the deep orange sweet potatoes typical of the American Southeast "dessert sweet potatoes." They don't like them.

"It doesn't matter if the yield is very good. If it doesn't taste good, then the farmers won't have it," Mwanga says.

Early on, Mwanga obtained some seed from the International Institute of Tropical Agriculture and began growing it on research center plots. As usual, word spread to the adjacent village that a new super variety of sweet potato was on the scene, and farmers began raids on germ plasm that constituted early, unofficial release of the variety. But the next year there was no problem whatsoever with theft, because farmers had grown the foreign sweet potatoes and had tasted the product.

Aside from making importation of germ plasm difficult, the preference for white sweet potatoes raises a conundrum that Mwanga will probably wrestle throughout his career. The orange color in other varieties of sweet potato indicates the presence of carotene, which means they are high in vitamin A. This linkage has worked to the benefit of breeders in places like the United States, in that higher nutritional value is tied to what most people there regard as better taste. The same preference has had most Americans eating more vitamin A in the past few years as a result of the introduction of "baby carrots," which show up already peeled and in plastic bags on supermarket shelves. They are not really young carrots at all but, rather, the core of a high-carotene carrot bred by Henry Munger of Cornell University. A specially built machine mills them down to "baby" size. Munger performed a similar trick with squash.

In Uganda, food preferences create an opposite effect, and children on their low-carotene diet suffer from a lack of vitamin A. The deficiency could be corrected in a few seasons if breeders could somehow magically decouple orange color and the associated sweetness from carotene—probably impossible. More likely, breeders will have to gradually increase the carotene in national varieties. This issue remained in the back of Mwanga's mind as he began to tackle the more pressing problems of resistance to weevils and virus. Toward that end, CIP had sent him fifty virus-resistant seedlings. "We put them out there and they were wiped out the first year," he says.

The imported stock seemed defenseless against the region's virus strains. At least, though, CIP had developed some protocols for breeding virus resistance. If he couldn't use the results, he could use the tools that created them. Under CIP guidelines, the first step was to make crosses of the various promising lines, then set out 3,000 to 10,000 seedlings of each progeny and see what happened. From there, it would be a simple matter of selecting the survivors—virus-resistant by definition—and pressing on. Mwanga began the process, but so great was Uganda's virus pressure that there were no survivors. None. Mwanga learned he needed to increase his initial set of seedlings tenfold over the standard used by the rest of the world—to set 100,000 seedlings, not 10,000, of each cross. That is, the stiff, tedious, backbreaking labor of plant breeding was multiplied by a factor of ten inside a research station in which decades of civil unrest had reduced the staff to near zero.

Nonetheless, the program identified and released six resistant cultivars in 1995. Five of these were landraces already growing in Uganda and blessed with some resistance that screening confirmed. One was a product of the first wave of breeding. None of the six is completely resistant, but they are an improvement, a foundation.

Experiments with international varieties suggest that the likelihood of finding solutions increases by looking close to home. Given the popularity of sweet potatoes in the countries surrounding Uganda, something might be gained through a regional strategy. Uganda had, in fact, begun pursuing such a strategy, cooperating with most of the countries in the immediate area. That effort formally sits under CIP's umbrella now as a French-named organization with the acronym PRAPACE. In the late eighties, Rwanda was a particularly strong player in the region. A close alliance between Uganda and Rwanda developed quickly and one between Mwanga and a colleague in Rwanda, George Ndamage. By then, Mwanga was well into a breeding program, simultaneously tackling development of resistance to the virus complex and to weevils. He had also obtained some germ plasm from Rwanda, which was already fighting the same battle.

In 1993, Mwanga and Ndamage began to prepare a joint application to McKnight, no easy task. At the time, the Namulonge Research Station had no working phone or fax machine, so every time Mwanga needed to communicate with his partner, he would bounce the hour's drive into Kampala to place a call. The final stages of the application process were best handled by E-mail, which couldn't be sent even from Kampala, so Mwanga flew to Nairobi, Kenya, where E-mail was available. In Uganda and throughout Africa, one often hears the phrase "It is impossible." At first it does not register in all its finality on can-do American ears. Then experience finally teaches one to hear it properly, as the irrevocable slamming of a steel door. "It is impossible," someone will announce, and he means it. And if he doesn't, the infrastructure does. After a time, even a visitor comes to accept that some seemingly simple matters are indeed impossible, or at least damned difficult.

In 1994, the Rwandan government turned its attention to varieties of humans, rather than plants, and adopted a policy that called for its Hutu majority to murder its Tutsi minority. The Hutus hacked at least 800,000 people to death in 100 days. Included in the carnage was an uncountable portion of the country's future, its intellectual power, a subset of which was the field knowledge hard-won by George Ndamage.

"He was at home," says Mwanga. "They took him and his children to his field, to his sweet-potato experimental plots, and they killed him there . . . They were ahead of us. I spent two weeks there. They had a very good program, but that is history now. [But] I got seed from them."

Rwanda lost its only breeder, but a plant pathologist there, Martin Bicamupaka, fled the country with the help of a regional network and went to work first in Nairobi, then in Uganda. Research continued. New strains of resistant sweet potatoes emerged, and a few years after the massacres, regional food experts, including USAID, believed the time was right to aid the struggling Rwandans. The organizations scheduled a meeting in Kigali, Rwanda's capital, and asked Bicamupaka to attend. His presence would send a strong signal that the region had overcome its political difficulties enough to get back to the urgent business of food. Bicamupaka, leery of returning, refused. Mwanaga says USAID officials persisted, finally persuading Bicamupaka to attend. While the meeting was in progress, someone came and asked Bicamupaka to step outside for a message.

"They came for him. They took him to jail, and he died there," says Mwanga.


A fundamental notion of the McKnight program holds that the real work is primarily local, driven by people working in their countries. However, when the Uganda sweet-potato project became one of the nine successful applicants, it was also at a point that the intellectual infrastructure of the crop had become dangerously narrow in Uganda. Something of a dilemma arose. Mwanga needed to complete work on his Ph.D., which would give him the advanced training in breeding techniques he would undoubtedly need to tackle the problems he faced. This meant his leaving the country, but his role was so pivotal that, in his absence, it would be difficult for the work to go on. The overseers of the McKnight project chose to sacrifice immediate progress on the project in favor of Mwanga's Ph.D. work in the United States at North Carolina State. He spent almost two years there, mostly training in molecular-marking techniques. He returned to Uganda early in 1998 for fieldwork that was to take a couple of months but dragged on to a year because of the complications of the field trials. After the stint in Uganda, he was to return briefly to the United States, then go to Lima, Peru, to conduct some research at CIP.

Beyond the logic of Mwanga's specific case, a somewhat deeper justification of this sort of globe-trotting research is buried here, demonstrating the way in which local work embeds itself in the international network of technology. That is, between the poles of isolation on one hand and hierarchical export of technology on the other, a more mature and intricate relationship between the developed world and the rest is evolving. Mwanga's is a case in point.

We are walking the ground of Namulonge as Mwanga explains his work, laid out in long screenhouses full of sweet potatoes in varying degrees of distress. On each of hundreds of plants, he meticulously grafts virus-infected material to screen for virus resistance. The grafts are the surest way to guarantee infection. A short walk away from the screenhouse we can see field trials, rows of sweet potatoes strung on trellises. Farmers don't use trellises, but the support exposes more surface area of each plant to pollination and to pests.

All these thousands of plants in the screenhouse and the field are part of field trials that last, typically, about eight years. Halfway through that period, resistant strains emerge, but it takes three or four more years to confirm that resistance in the field, to prove what the breeder already knows. Meanwhile, all around Namulonge, stretching clear to the country's borders and beyond, an endless string of small plots exists, each vitally dependent on sweet-potato yield.

A shortcut involves the identification of a molecular marker that would allow researchers to pinpoint the section of DNA that confers resistance to the virus. Confirming resistance by molecular markers obviates years of field trials. That technique, however, requires sophisticated facilities for conducting molecular biology. At Namulonge the toilets often don't flush because the water system fails. If one is meeting with someone of sufficient importance to have an office telephone, the meeting will be interrupted for any phone calls, because the phones work so infrequently that people take calls when they can. One lab uses a radiotelephone to circumvent the whole system. They do have E-mail now, but the several dozen people at the station all have the same address. The whole station's E-mail is downloaded, printed out, and distributed once a day.

The station is simply not equipped for the molecular biology Mwanga needs to do. Nor can the problem be solved by simply buying sophisticated equipment when the whole system from power lines to water lines can't support it. So instead he travels and does part of his work in the United States and Peru. Necessity has mothered invention, not just in Mwanga's program, but throughout Uganda's network for agricultural research. He might seem alone in his program, but, in fact, he is part of a consistent national strategy, but one player in a nationwide umbrella, the National Agricultural Research Organization. After completing his undergraduate work, he took his first job and wanted to do research on ponds. The managers of the country's research network, however, steered him toward his career as a plant breeder, providing each new notch of training he needed.

Theresa Sengooba is one of the people important enough to have a phone in her office. As the director of Namulonge, she is responsible for tying the station's research in five crops to the needs of the nation.

"It's not like in a university where you see an interesting problem so you start to work," she says. "How all this research gets translated to the improvement of agriculture, that's the challenge we have."

This goal may seem obvious, but there is a deeper layer to the logic that sets Uganda on a clear and distinct course in its work. In the early nineties Uganda went through a planning process complete with help from consultants and World Bank money. The intention was to develop a lasting framework for research, production, and processing. Initially, the work concentrated on cash crops aiming toward a classic model of development bootstrapped on agricultural exports that would supply the capital for textbook Western-style modernization. This idea, however, was not new to the Ugandans.

Mwanga recalls that during the darker days of Idi Amin's rule, the dictator grew increasingly dependent on infusions of foreign cash. Amin's troops distributed cotton seed through the villages and ordered farmers to grow this cash crop for export. Mwanga says the villagers surreptitiously roasted the cotton seeds, then planted them under the soldiers' watch and pretended to share the soldiers' puzzlement when the government seeds would not germinate.

The Ugandans who had been part of the planning process in the early nineties remembered such lessons and began challenging the assumption that export income and development were better than food security. They decided to reverse priorities and to pay attention to cash crops only after building a strong base for subsistence agriculture. Instead of cotton and coffee, they concentrated on sweet potatoes, cassava, and bananas, a decision that generated a very different-looking structure and set of problems. Research on diseases, pests, and breeding was still necessary, but socioeconomic concerns focused less on generating capital for building processing plants than on seeing whether there were enough trucks and roads to get the sweet potatoes to local markets.

It also generated a set of needs unique to Uganda. For instance, the country could no longer import an off-the-rack, one-size-fits-all cotton infrastructure aimed at punching out finished textiles. The issues affecting culture and distribution of sweet potatoes, cassava, and bananas are primarily unique to the region, particularly given the specific social and natural setting of Uganda. Instead of dealing with a few managers of mammoth corporate farms, the planners needed to build a network that would encourage a two-way flow of information with poor farmers. It also implies an attitude toward technology that Gadi Gumisiriza, the nation's principal agricultural research officer, characterizes as "shopping."

Mwanga and I are sitting in Gumisiriza's office (it has a phone) in another aging sprawl of colonial buildings near Entebbe, the nation's old capital, which is about 30 kilometers from Kampala. Mwanga raises the problem of the foreign sweet potatoes that Ugandans won't eat. Gumisiriza himself is a plant breeder, with a Ph.D. he earned in London and a background in cereals. He has similar tales from his own experience.

Gumisiriza argues that his country will indeed need to import a heavy dose of technology from the developed world but, again, re-shape it to better fit the local situation. The key to all this, he says, is a corps of national scientists who can pair a cosmopolitan education with knowledge honed through contact with local field conditions. These people will be the appropriate conduits of technology.

"We have people shopping around the world to see what technology can be used here," he says. "If there is something that will work, we'll pick it up and use it."

Mwanga focuses on viruses now, but the weevil war continues. Down the hall from Mwanga's office, three entomologists huddle over a chopped-up five-liter plastic jug, once the great hope of the weevil war. Two species of weevils create the bulk of the problems, and both behave pretty much alike. The adults, small flying insects, lay eggs that hatch larvae, which bore into the roots of the sweet potato to feed and mature. At harvest time, farmers dig up roots riddled with the tunnels of the larvae. Yield losses are heavy and almost total in some areas.

In part, the problem could be dealt with by timing the harvest. Infestations spread rapidly just as the crop reaches maturity; weevils, like humans, are fond of ripe sweet potatoes. If farmers harvested promptly, they could avoid the bulk of the infestation, but that is precisely when all the sweet potatoes in a region are ripe, so the market is flooded and the prices are lowest. Ripe sweet potatoes will keep well unharvested for several months, so harvest is timed not so much by the fact that the tubers are ripe as by the price. Thus every farmer faces a gamble between leaving his crop in the ground to be ravaged by the weevils and taking it to market early to be ravaged by the law of supply and demand.

Ugandan researchers are keenly aware of the significance of all this and had some hopes that pheromone traps, which have been tried with great success in the southeastern U.S., might be the magic bullet. Pheromones are the chemical signals by which insects communicate, and those issued during mating season powerfully attract bugs of the opposite sex. Pheromone traps use a small amount of a female's attractant in a simple device that lures and ensnares so many males that breeding success drops dramatically. A typical trap in the United States can doom 10,000 males in a single evening. No males, no breeding, no larvae—so no tunnels in the tubers. Besides being effective, the system has obvious advantages over the use of insecticides.

The Ugandan entomologists Benson Odongo and Paul Laboke wanted to ensure that their pheromone traps were examples of appropriate technology. Instead of simply appropriating the design of a trap from the United States, the researchers traveled to a series of Ugandan villages to find out what sort of plastic jugs were readily available as trash or cheap water jugs. They then experimented with methods of carving the local jugs to determine which was most effective and which could most easily be modified by the farmers. They experimented with different sorts of vials for containing and distributing whiffs of the pheromone from within the traps and eventually found that two types were necessary, a rubber one for one species and one of polyethylene for another. They made sure a ready and cheap source of both vials and pheromones was available, and they came up with a simple and harmless solution of soap and water that would kill the hapless males lured to the traps.

When they went to test plots, the traps worked. They began trapping males from the first day of planting, not in great numbers—nothing approaching the 10,000 a night common in the United

States—but periodic censusing indicated a remaining population that was close to entirely female. However, at harvest time there was no payoff. The test plots of sweet potatoes showed as great a loss to weevils as did a normal field. The researchers were missing something. Perhaps weevils were ranging farther than they thought, or perhaps there'd been a timely introduction of a gang of males from the next plot. Repeated trials showed no improvement in yield.

So they took a close look at the weevil itself and got an answer. Unlike with sweet-potato weevils in the United States, the females of the two species causing all the grief in Uganda need to mate only once during the entire breeding season. During this encounter, they take on enough sperm cells to supply a normal year's worth of eggs. Further, one male can service twelve females on average. The pheromone traps were simply ineffective in the face of such tremendous copulative efficiency.

Odongo and Laboke tell this story almost as if it is a little joke the world played on them, although it can't be terribly pleasant to have so many years of research come to a dead end, especially when so much hangs on one's work. But scientists to the core, they remind me that negative information is still valuable information. Besides, now they have a trap that works. It may not control weevil damage, but it traps weevils for censusing. Nor have they given up. They are using the traps to see if they can disrupt mating in some other way.

Sam Kaabi, a grad student and the third entomologist working with the project, joins our conversation. He has carried out a small bit of work that points toward some leverage on the problem. It turns out that various varieties of sweet potatoes have different inherent rooting depths, the point below the surface where the tuber develops. According to Kaabi's research, a correlation exists between rooting depth and weevil damage. That is, a genetic characteristic indirectly confers weevil resistance but does not necessarily lock farmers into growing this variety, nor does it lock breeders into selecting for the trait. Instead, it says simply that weevils lay their eggs in the portion of the tuber easiest to reach. The deeper the tuber, the safer.

Another bit of information figures in all this. Yield loss is greatest in the driest regions of the country, and that loss explodes with the onset of the dry season. This does not so much signal the weevils' preference for arid conditions as it says something about the soil. As it dries, it cracks, and those cracks give the weevils more access to the sweet potatoes. The entomologists have found that by simply instructing farmers to till and mound the soil around sweet potatoes as the dry season sets in, they can dramatically cut losses to the weevils.

This is the sort of solution that is not particularly dramatic but is typical of the range of small, simple steps that add up to integrated pest management. The method, almost by definition, abandons the search for magic bullets. It is, in fact, a reaction to the most highly touted magic bullet of the industrial age of agriculture, the widespread use of pesticides, a solution that worked like squeezing a balloon. The pesticides were dramatically effective in a given area, but their use failed to take into account the broader system in which they operated. Not only do they endanger the health of agriculture and consumers, they also kill predator insects and force an unnatural selection that develops pesticide-resistant strains.

It has been said that there are three stages of agriculture, and only the last is sustainable. The first is characterized by the industrial model and the use of a single tool, such as a pesticide. In the second stage, we substitute less damaging replacements for those tools— pheromone traps in place of insecticides, green manure to replace nitrogen fertilizer, drip irrigation to replace inefficient flood irrigation. Only in the third stage is agriculture incorporated within the system in which it operates to take advantage of what we are coming to call "ecosystem services." The final step embeds solutions within the advantages and limits of natural and human culture.

The entomologists at Namulonge ended their story by laying out solutions to the weevil problem, almost reluctantly, apologetically. They had no sweeping scientific discoveries to offer, only a series of incremental solutions that had to be incorporated into the culture through painstaking farmer-to-farmer communication. Hilling the sweet potatoes will help. So will early-maturing varieties that allow some farmers to harvest before crop prices drop. So will development of cheap storage structures that allow farmers to harvest at the optimum point but to hold their produce until prices rise or until it is needed for household food. So would a better system of roads; farmers are now at the mercy of truckers and must sell when they show up in the village to buy. So would a range of cultural practices. For instance, instead of discarding infected portions of tubers in the field, which allows the larvae to mature and to return as adults the following season, researchers are urging farmers to keep pigs to eat the infected tubers, converting a source of infestation to available protein. Larvae can survive also in discarded stems, which are also normally left in the field to rot. Instead, the farmers could dry and burn the stems.

All these steps are small but get to the root of long-established practices in Uganda. A set of ingrained practices we call culture, the root of agriculture. Culture changes only slowly, imperceptibly, incrementally.

The woman's farm grows beans, peanuts, maize, bananas, cassava, sweet potatoes, and a few coffee trees for cash. She has some goats for milking, but nowhere for them to graze on her plot. She doesn't know how big her farm is. Robert Mwanga smiled before translating my question: "This will be a problem. She won't know."

It looks smaller than a couple of acres. Her children herd the goats at village roadsides to browse and cut elephant grass to feed them. She has a tiny house made of the local crude bricks. There appears to be no husband, at least not resident, but she says she has one. I ask if he helps with the farming. Sometimes, she says. Sometimes he gives her money to help plant sweet potatoes. She suffers losses to virus and weevils, especially because the crop of sweet potatoes, now ripe in their neatly hoed hills, remains in the ground until the family is ready to eat them. Mwanga says they also act as a sort of bank. When the household runs out of kerosene or salt or some other staple, she will dig some up and pile them at the roadside to sell for a few pennies.

Add a few other crops to the mix: mango, jackfruit, avocado, sugarcane. Then wind the kilometer on kilometer of hard-packed clay roads through Uganda's verdant hills. Six of us sit in a Toyota pickup truck whose shock absorbers were long since beaten to death. We sit four across in the back seat of the extended cab, so that a set of shoulders protrudes from each window, each set often simultaneously slapped by limbs of trees crowding the road that is more bike path than truck path. Now and again our driver manages to find a particularly distinguished specimen of pothole, and four heads in the back seat smack the roof in unison. The driver cackles maniacally, and we jostle on down the road.

Sometimes something stands out from the broad pattern—for instance, a house that seems a mansion against the patchwork of huts. A sign explains that the local herbalist resides there, a doctor of sorts, a practitioner of traditional medicine or a quack, depending on one's point of view. Then we meet two farmers, Stephen Mutale Lutuna and Bisaso Simbun Samuel. In their late twenties, they are well-spoken and engaged, and take us to fields of sweet potatoes, rented plots. They are not so interested in me as they are focused on the rest of our group. They have learned that the men with me are sweet-potato scientists, so they fire question after question at them, while the scientists learn what they can from the young farmers.

I interrupt to ask the men how they became so well educated for farming, and they laugh at me. They are educated, but not in agriculture, and neither intended to be a farmer. Each imagined something a bit closer to an office with a phone and a shirt with a tie. Uganda provided them no opportunities to pursue those aspirations, even with a decent education, but it lets a person farm, so they rented a bit of ground and they farm sweet potatoes for cash. They do it with curiosity and commitment and a recognition that these scientists might offer something to advance their enterprise.

The weevil problems have been terrible. What should they do? Should they apply Ambush (an insecticide)? No, don't do that. The scientists tell them what the Ambush vendors won't: it doesn't work against weevils, because they do not feed on the surface of the plants, where the insecticide is sprayed, but deep within the tuber, where it cannot reach. The scientists explain the trick about hilling the sweet potatoes when the ground cracks and about the necessity of a timely harvest.

What about varieties? The farmers have heard that there are scientists at a place nearby who are trying to breed special sweet potatoes that do not succumb to virus, but they have also heard this variety does not work in real farmers' fields.

Mwanga points to a particular hill of sweet potatoes and asks the men about it. Is it good? Does it weather the virus? Oh yes, they say. They have been using it for a couple of years and it has a name, which they pronounce. The name is strange to Mwanga and he asks about it. They tell us it is named after the local woman, a farmer, who brought it into the village a few years ago.

Mwanga laughs. The variety in question is one of his, bred in the early days of the Namulonge project. He doesn't care what they call it, as long as it works, but admits that he did find one village where farmers named it after him.

Finally the talk boils down to yield, which keenly interests these young men, as it does all farmers. They get about six tons to the hectare. The researchers tell them they could get at least triple that, and the men get excited. They make plans to visit Namulonge, which, on these roads, is more than an hour away.

Farther on, the roads seem to get even worse, until at the end of an all but impassable stretch the way opens to a larger farm with a big house—a masonry block house that would make a two-bedroom bungalow in the developed world—standing on the farm's highest point. The farmer is an old man with a white-fuzzed face, Edward W.S. Kamya Lugonvu, and he invites us to sit in his living room, ostensibly so that we may sign his guest book, but also so that we can observe that he owns a working television. He watches CNN via his satellite.

This farmer is, by all measures, prosperous. He knows how much land he has (ten hectares), but when I compliment him on his farm, he corrects me. It is, alas, only a bit of a garden; were it a proper farm, it would have a tractor.

"I don't have anything other than this hoe," he says. "We should have a tractor."

He knows his business cold, walking the touring group through each plot of each crop—maize, coffee, sweet potatoes—and explaining in great detail his strategy, his costs, his methods, and the reasoning behind his methods. He shows us a field of maize laid in grids that is a test plot he set up to check the advantages of a new planting system. He explains how he measures the yield from this plot to optimize and best use the results.

Mwanga says the old man is in the business of filling institutional contracts, supplying produce to prisons, schools, and hospitals. He has done well at it, but old Edward doesn't want to talk about farming so much as he wants to show us a corner of his land that is not farmed. He rushes through the tour to get us to a bit of rain forest next to a small stream.

He says he used to like to go to Entebbe because there was a botanical garden there, to travel the 40 kilometers or so on a weekend outing and sit and look at the trees. Then he got to thinking about the few trees on the corner of his farm. What did Entebbe have that his farm did not? So he hired some help and cleared out the understory, leaving maybe an acre of tall trees arching over his bit of stream, which he has dammed to make a small pond barely big enough to float the small rowboat he hopes to obtain.

"This is my lake to compete with Entebbe," he says. One must use considerable imagination to see it dwarfing Lake Victoria, but Edward has considerable imagination.

"In the village here, the people have nowhere to go. A poor man doesn't have a car to go to Entebbe to see the trees, so I built a botanical garden so our people don't suffer to go to Entebbe and spend a lot of money along the way."

It may not be much, but it is at least a peaceful corner of the world in a village that not so long ago saw slaughter.

In one of our interviews, Robert Mwanga had been walking through the long list of obstacles to his work specifically and to his country's growth in general. The obstacles now seem huge, but on the other hand, he said, the wave of violence has passed, at least for now.

"One thing we appreciate," he says, "is peace."

# From Basket Case to Bread Basket

## When Biotechnology Has a Brain Trust

### [India]

A small room in a palace at the edge of Pune, India, was once a cell when the palace was a prison. Imamsultan Muhammad Sha Aga Khan III built the palace in 1892. It still stands straight and well kept, bleached to a pale yellow in the glare of a south Indian dry-season sun. On a Sunday afternoon a few visitors wind through the halls and great rooms, now a museum. More quietly, they walk along a veranda at one edge of the palace. It fronts a series of small rooms that held for a time Mahatma Gandhi; his wife, Kasturba; and his secretary, Mahader Desai. All three were imprisoned here by the British for nearly two years after Gandhi galvanized his country's independence movement with his "Quit India" resolution delivered in 1942. Both Kasturba and Desai died here; their ashes remain on the grounds. Otherwise, there are only a few artifacts. Gandhi shunned possessions, so what does remain is all the more significant. A visitor can pause a long time to look at the pair of sandals left in the room he once occupied, or his spinning wheel, which would become the symbol of this man. In the hive-like markets of Pune, a city of 2.8 million, one still finds whole shops dedicated to the sale of khaddar cloth, the handspun cotton material Gandhi championed. Before Independence, India grew cotton for British mills, a servitude the

British reinforced by demanding that India buy finished cotton goods back from the British. The handspun cloth was Gandhi's defiance of this system, a sort of wearable flag of independence that flies to this day.

At the opposite edge of the city, near Pune University (the oldest in India, called the "Oxford of the East"), sits another set of sun-bleached buildings on a sort of campus. Sprawling to about 600 acres, the National Chemical Laboratory is the size of a suburban university. Houses and flats for the staff ring a series of stone, marble, and masonry buildings and meticulously tended flower gardens, all of it anchored by an imposing main building with a cornerstone dated 1947, as good a date as any for the cornerstone of India's national science. Scientific research became institutionalized at Independence.

This building housed a precursor to the National Chemical Laboratory. The latter began in 1950, along with a network of research institutes, the legacy of Gandhi but especially of his successor, Jawaharlal Nehru. From the beginning, the impetus for research was primarily practical, a drive to alleviate poverty by applying the techniques of modern science. According to a participating scientist, it began largely by failing.

D. R. Bapat, a retired plant breeder, sits in a conference room at the National Chemical Laboratory riffling a pile of papers from which he periodically fires off statistics about chickpeas, his latest passion. He's had many passions in a diverse career that eventually brought him to the position of director of research at Mahatma Phule Krishi Vidyapeeth, the state of Maharashtra's leading agricultural college at Rahuri. Bapat made his early mark in sorghum, a generation ago breeding the hybrids that played a key role in saving the nation from what everyone then expected to be generations of famine.

By the time of Independence, India was well aware that its vulnerabilities were not all limited to imported cotton clothing. It was also already importing large amounts of rice, wheat, and maize. Institutions like the National Chemical Laboratory and the careers of scientists like Bapat were founded on the goal of ending those imports. The country began buying fertilizer and irrigation equipment designed to modernize its agricultural infrastructure. The results, says Bapat, were effectively zero—zero for more than fifteen long, hard years, during which dependence on imported grains continued to rise. Importing technology simply did not work, because it had not been customized to fit the specifics of India's climate, markets, knowledge, and available technology.

Then, in 1964, the Green Revolution took off in India. Norman Borlaug's gains in Mexico spread quickly around the world, but especially to India, making those two countries the primary fronts in the food war. The Rockefeller Foundation helped introduce new hybrids of maize to India that effectively doubled yields. The technique spread quickly to rice and wheat, primary wet-season crops, with impressive results. The same techniques spread to sorghum, but not as effectively. The Green Revolution can be summed up in two words: short plants. By taking advantage of dwarfing genes, breeders encouraged plants to put the larger share of their energy into seeds, astronomically boosting yields. The problem was, Indian farmers used sorghum stalks to feed livestock and weren't especially anxious to sacrifice that forage to higher grain yields. Bapat's task was to use the developed world's dwarf sorghums to breed a compromise hybrid for India. He did so, bringing the crop in line with the other key grains, which collectively tripled yields during India's phase of the Green Revolution. The result is that India no longer imports its primary grains. It feeds itself.

So where's the problem? The question sets Bapat to thumping away at his pile of chickpea statistics and to ultimately laying out, in quick, crisp sentences, the fundamental case built on the paradox that the successes of the Green Revolution have, in turn, created great need.

In almost any discussion with an Indian about agriculture, the statement "We are primarily a vegetarian country" invariably makes an early appearance. It is this attribute coupled with Green Revolution gains that allows India to feed its masses, a fundamental efficiency that makes it a model for a crowded world. Put simply, it takes roughly one-tenth the resources—sun, water, land, fertilizer, and so on—to produce a gram of vegetable protein, compared to a gram of meat protein.

The Green Revolution targeted grains: wheat, maize, and rice. These are sources primarily of carbohydrates, not of protein. India's main sources of plant protein are pulses: chickpeas, pigeon peas, and lentils. While production of the big three cereals tripled in India between 1950 and 1998, yields of chickpeas rose, per hectare, only from .4 ton to .5. Even this understates the problem, because the newfound success of the primary cereals enticed commercial farmers to grow more of them, pushing pulses off the land. The minuscule gain in yields was offset by this second factor, leading to a net decline in overall production. The results can be read in the diet of the average Indian. A person needs about 80 grams of plant protein each day. In 1961, the average Indian was getting about 70, already a deficient diet. A generation after the Green Revolution, declining production coupled with a greatly increased population yielded only about 37 grams, half what is needed.

The problem has a mirror image in the health of the land. The bulk of India's subsistence agriculture is built on at least two seasons of cultivation. There is a wet-season crop of rice or wheat and then a dry-season crop of sorghum, chickpeas, lentils, or pigeon peas, all of which are drought-tolerant. In addition, the pulses are nitrogen fixers, taking free nitrogen from the air and leaving it behind in the soil. Bapat says chickpeas typically leave behind 30 to 35 kilograms of nitrogen per hectare, supplying about half the fertility needed by the grain crop that follows. Also, pulses secrete phosphorus from their roots, which the cereal crops also need. The emphasis on grain yields compromises fertility, ultimately undermining grain production as well.

"The graph of yields is plateauing, and in some cases it is also showing downward trends," Bapat says. "We have to bring in a second Green Revolution. This is how we should think of it."

Meanwhile, there are some numbers that suggest a deeper layer to this problem: using existing technologies, France gets 5 tons per hectare of chickpeas, ten times India's yield. England gets 3 tons.

Three middle-aged women in saris huddle around a tape recorder (each has been introduced to me with the uniquely Indian honorific of "Dr. Mrs."): Vasanti V. Deshpande, a biochemist; Vidya S. Gupta, a molecular biologist; and Mohini N. Jainani, also a biochemist. Together with P. K. Ranjekar, the genial man who heads the National Chemical Laboratory's Division of Biochemical Sciences, they front the brain trust that the laboratory has pitched against the protein problem. The national laboratory is joined by the agricultural school, Mahatma Phule Krishi Vidyapeeth, which carries out the plant breeding and fieldwork, perhaps later some of the molecular biology as well.

The team is broad-based, but it seems tight. Their discussions have the quality of thinking out loud. One scientist frequently finishes another's sentence. Each seems fully conversant in the details of the others' specialties.

They are the fruit of their country's established tradition of institutionalized research, which makes India an exception in the developing world and a happy hunting ground for developed-world corporations hoping to recruit their way out of the educational inadequacies of places like the United States. India has produced an intellectual infrastructure good enough to be raided.

At the National Chemical Laboratory I met a long string of young scientists all working on chickpeas, Ranjekar's team. In the twenty-two years he has been at the laboratory, his group has turned out thirty Ph.D.s, themselves a product of a systematic national strategy for training and recruiting scientists. Each year, 25,000 candidates to enter doctoral training sit for nationwide tests that will decide who is awarded fellowships for the rest of their education. There are slots for only five hundred. Because of its reputation, the national laboratory is able to attract the top finishers from among the five hundred, each year signing on a couple to do their Ph.D. work at the laboratory for five or six years. Because the system is based completely on merit, it pulls in a cross section of students from the country, which is overwhelmingly (about 80 percent) rural. As a result, many of the biochemists and molecular biologists in Ranjekar's lab speak firsthand and with passion about the conditions of farming and rural village life.

As do the scientists. "Last year we had such terrible losses because of helicoverpa that many farmers in this region committed suicide," Gupta says. "There were five hundred suicides across the country."

Despande finishes the thought: "One day the farmers are very happy that God has been good and the rains are on time and the field is full of crop, and the next day when they harvest, the yield is very low."

*Helicoverpa*, the bug of this story, is more commonly known as the pod borer. The adults lay eggs on the chickpea plants. Those hatch and more or less languish on the plants as virtually invisible hair-sized larvae, until they hit a growth stage. Just when the larvae demand extra food, the chickpeas flower and set pods. The larvae bore nearly invisible holes through the pod wall, burrow inside, and eat the developing peas, then move on to the next pod, typically gutting four or five pods a day. At harvest, a farmer goes to a field of apparently healthy, robust plants only to find the pods hollow. Losses in bad years run to almost total.

The insect can be controlled to a certain extent by insecticides, but their cost puts them out of reach for many farmers. Nor are the Indians particularly anxious to follow that path. They tend to be somewhat fastidious and health-conscious about their food, in the first place, and they are becoming increasingly conscious of the mounting problem of insecticide resistance caused by overuse of chemicals. They are searching for neater ways to deal with pests, like the field full of common egrets I saw systematically pecking away at pod borers. "Biological control," they said.

Apparently no varieties of chickpeas have inherent resistance to pod borers, at least not now. Chickpeas originated in the Middle East but traveled trade routes to India long before people were recording such movements, so they have probably been in India as long as agriculture has. They've taken a firmer hold there than anywhere else, in that India accounts for about 70 percent of the world's total production. The country mostly produces a smaller, smooth-seeded group of varieties known as *desi*. Westerners tend to picture the larger wrinkled seeds of the kabuli group of varieties, the garbanzo beans common through Europe and the Middle East. There is much variation in the species, but unfortunately this diversity fails to yield resistance to pod borers.

Other species of crops are similarly vulnerable. The pod borer is polyphagous, which means it eats a lot of different plants. It eats something like one hundred different species, settling, during the off-season for chickpeas, on hosts such as cotton, tomatoes, and cabbage. To the agronomist, this means the insect problem is atypical, not an artifact of monoculture and therefore not controllable with cultural practices such as rotation and intercropping. The enemy here is a generalist, downright human in its ability to adapt and persist using its diverse appetites. To the biochemist, this suggests an even more complex problem. Natural immunities of plants are generally based on a pest's inability to digest certain substances, and this bug has already shown itself to have a gut proficient at grinding up most anything. But not everything. The pod borer cannot invade some plants, and that is what sent scientists looking for protease inhibitors.

Protease inhibitors are well known for their spectacular role in treating AIDS, but they are actually simply proteins that prevent certain enzymes from working. Some enzymes work at food digestion.

The pioneering research in the field relevant to agriculture was done by Clarence Ryan of Washington State University almost thirty years ago, when he first demonstrated the way plants use protease inhibitors to ward off pests. He found the chemicals especially abundant in legumes, which include chickpeas. During a trip to the United States in the early nineties, Ranjekar called on Ryan, a visit that led to a plan and a partnership. Washington State allied itself with Ranjekar's group in order to secure a grant from McKnight.

The researchers went looking for protease inhibitors in chickpeas, but at first found none—no real surprise. Chickpeas seemed defenseless. The search, however, went on in an effort to find some vestige of inhibitors at some isolated place and time in the plant, some latent trait that could be amplified or exploited to provide a defense. Up until then the group had been using relatively straightforward assays for the inhibitors, but decided to try instead a new method that involves a gel X-ray contact print. This new window revealed that chickpeas produce seven different types of protease inhibitors. Further, they produce them right in the developing seed where they were most needed. Defenseless? Hardly. The task, then, was to probe the general-purpose gut of the pod borer to see what allowed it to conquer these defenses. A protease inhibitor is still a protein, simply one designed to defeat and disable the digestive enzymes of the pod borer. That may have worked once, but the pod borer has adapted by producing a series of new enzymes that not only overcome the defense but digest the inhibitors.

"The insect protease is more powerful than the chickpea. The pod borer is not only insensitive . . . It is feeding on them. It is using them to its advantage," said Deshpande.

Their method was like an attack that could not only shoot down defensive missiles but drain them of their fuel to power the offense. The researchers were discovering the history of an evolutionary arms race between plant and pest with a series of at least seven escalations and counterescalations. The current round is being won by the pod borer. So much for the notion that resistance arose with the use of artificial pesticides.

Still, there are plants—the common groundnut, or peanut, and the winged bean (a common Asian legume)—that the pod borer cannot eat. In these plants researchers found a series of new protease inhibitors that could stop the pod borer. At this point readers who have been following the headlines will know where this discussion is headed, especially since there are molecular biologists at the table as well as biochemists.

At about the time the Indians were telling me their story, the Japanese were announcing moves to ban transgenic crops. The British Isles were well into a raging controversy about "Frankenfoods." An international conference convened in South America to design protocols for exports of transgenics blew apart in a rancorous huff. Health-food stores in the United States were mounting an effort to purge their shelves of transgenics by denying them organic labeling. *The New York Times* was editorializing against the use of transgenic maize.

The headlines conjure pictures of what a transgenic might look like—square tomatoes, Monsanto's factory farm seeds built like machines, a sort of Erector set of food. None of the images seemed to capture the story that was being spun around that table by three matronly Hindu women in saris. Still, their story concerns transgenics.

Remember that protease inhibitors are proteins, and genes do their work by forming proteins. That's how the genetic code is, in the apt language of the science, "expressed." To arm the chickpea, researchers plan to transfer the gene from peanuts. Once included in the code, it will breed progeny that pod borers can't eat, or at least will have a very difficult time eating.

To those spooked by such methods, the researchers counter with the argument that they are merely teaching one food plant's trick to another, so the relevant chemistry is something humans already eat and, as the pod borer has learned to do with other inhibitors, already convert to our use with the help of cooking and enzymes. For those who believe we are crossing some godlike threshold by tampering with the genetic code, the researchers point to the evidence inherent in the chickpea itself—that it has already crossed this threshold at least seven times. That is, simple mutations, natural shifts in genetic structure, created new protease inhibitors that natural selection spread through the chickpea population long ago. By this light, the deliberate transfer of genes is only a directed mutation.

Over the course of the next few chapters, we will mention other examples of the use of genetic engineering, some more dramatic and alarming. These will be the basis for a broader consideration of this issue toward the book's close. And consider it we must, because genetic engineering is not speculation; it is already a fact of life in the agriculture of the developing world. For the moment, though, let the subject rest with a thought from Dr. Bapat, a lifelong expert in plant breeding, the discipline that is to genetic engineering what the typewriter is to the desktop computer: "The yield will jump to one ton without doing anything. The farmer has not to spray anything, has not to dust anything."

He need only plant a new seed.


R. B. Deshmukh is a plant breeder, as well as the director of research at Mahatma Phule, the job Bapat held until he retired. There must be some status attached, because our group has gathered an entourage, has formed a bit of a parade actually, down the few kilometers of dusty road from the campus. Deshmukh and I ride in a velour-upholstered sedan with curtained windows that looks like something straight out of the Raj but is actually a recent manufacture of Hindustan Motors, the national motor company whose tenacity regarding long-established design has dotted India with antique-looking cars. They are the automotive equivalent of khaddar cloth.

Behind our rumbling sedan trails a line of rattling compacts and small pickup trucks, altogether bearing maybe fifteen agronomists, bug guys, and breeders. We pile out of the cars at the edge of a field of chickpeas. At the far side of the field stand two small stone buildings the size of sheds: Hindu temples.

Deshmukh walks the grids of the chickpea test plots near the village of Rahuri, strides them really, with a pride that seems paternal. They resemble test plots at any of the world's ag schools, well-ordered and tended tiles of plants, labeled and bordered. If work is going well, one sees a dramatic display of differences, with some plants shriveled, withered, and dying, some just getting by, some thriving—evidence of differences a plant breeder exploits. Deshmukh can legitimately claim some paternity over this field; he has developed most of what is on display, beginning in 1975. India's systematic effort to improve the line actually dates back about ninety years. A variety that was released in the 1940s is still growing, as are some from the sixties and seventies—but Deshmukh says chickpea research more or less broke off in the late sixties when all the attention shifted to cereal crops—until he revived it.

In 1975 Deshmukh began collecting and evaluating material from around the country. Then in 1982 he released a variety called *vikas*. In the regional language, Marathi, the word means "progress." Three years later came *vishwas*, which means "faith." Both considerably improved some traits like productivity and quality of the seed, but what was needed was resistance, especially to Fusarium wilt.

*Fusarium oxysporum* is a fungus that lives in the soil throughout peninsular India, the most important chickpea-growing region. It develops from spores that can survive dormant in the soil for six years, then produce fungal roots, or mycelia, that invade the roots of nearby chickpeas, sucking nourishment for themselves and blocking the chickpea's own conductive tissues. The host plant withers and often dies. In a good year, crop losses to the fungus are 10 percent; in a bad year—and there are often bad years—the loss is total.

Unlike the initial situation with the pod borer, however, some local varieties of chickpea showed signs of winning the arms race with the fungus, some inherent resistance that a breeder might exploit, and Deshmukh did so. In 1993 he completed development of the variety *vijay*, which means "victory" in Marathi. That particular entry capped a series of Fusarium-resistant varieties now in the hands of India's farmers.

"This," says Deshmukh, indicating the thriving plants alongside the wilted ones in the test plots, "was a victory against disease and drought."

Case closed? It would be nice to think so, and we'll indulge that fantasy first with a visit to a nearby farm. Dr. Dattatray Sahadeo Wane is by no means a typical Indian farmer. First, his farm is large by Indian standards, almost nine hectares (just more than twenty acres). It is highly developed, with up-to-date irrigation systems and tillage practices. It is commercial as opposed to subsistence: he grows crops for sale. He is largely dependent on hired labor, not his own, and is a medical doctor by training. He is also atypical by commercial farming standards in that he grows chickpeas as a cash crop and turns a profit. Most commercial farmers stick with the high-profit cereals and leave the pulses and legumes to subsistence farmers who can't grow anything else.

Clearly, though, Wane has not miscalculated. Our HM sedan lumbers into his yard trailing our entourage. Wane, already afield with a full flurry of notebooks, punches away maniacally at a calculator. He barely makes it through the cursory introductions to the visiting journalist before he has Deshmukh and leading members of the entourage buttonholed for an animated exchange of notes in heavily inflected Marathi. The subject, a translator tells me, is the timing and duration of irrigations, a subject that Wane seems to wax on about with manic enthusiasm. While he speaks, he uses a bright yellow highlighter to illuminate sections of his notebooks. I leave the scientists to it.

I watch a crew in a nearby field harvesting sugarcane, whacking it by hand and piling the fresh cane atop high carts drawn by sleek, placid white bulls with blue-painted horns arcing high over their heads. The bulls recline in the shade of the carts, ruminate, and watch the people work. A couple of researchers from Ranjekar's staff at the National Chemical Laboratory join me, and we walk Wane's chickpeas. Before the two men were Ph.D. students, they were both village boys, and they know chickpeas. But Ashok Giri and Abhay Harsulkar don't start poking into pods here, as is their usual habit afield. They seem a bit stunned, like kids in a candy store. They keep using the word "beautiful," modulated in a manner one usually associates with mountain vistas and national parks. Wane's chickpeas *are* beautiful: robust, lush plants with pods mostly unburdened with bugs. He will probably take something above his average yield, which is three tons per hectare, six times the national average.

I ask Giri and Harsulkar to project what they see here onto the poor villages they know. They tell me that the yield may be beyond the imagination of the average farmer, but the techniques to achieve it are not. This success is based on the new resistant varieties, which are available and affordable. Wane has controlled pod borer with a couple of closely monitored applications of insecticides, and those are already being used in the villages, albeit in far greater doses than is necessary. Wane has irrigation, but there is a government program for financing irrigation, and besides, irrigation is not that critical for this drought-tolerant crop.

The real difference between Wane's farm and those of poor villagers is Wane's notebook. He records every detail of his farming, because he understands that the information he learns in the field, combined with what he gleans from researchers and others, boosts his success. That information links him with advancing technology. He's educated in science or, more basically, in laws of cause and effect, a simple matter than gives him an enormous advantage over almost all his neighbors.

We're invited to tea in Wane's modest farmhouse. He takes the visiting writer and the plant breeder aside, and around each of our necks places a garland of showy tropical flowers. He hands me a bouquet, then from a tray he takes a dollop of red color and smears a short mark down the center of my forehead, a *tilak*, a mark for an honored guest in an Indian household. Then, in the smear of the *tilak*, he embeds a few grains of rice.

Giri, Harsulkar, and I had toured another university test farm of chickpeas, but the fields are no longer being used for tests. Instead, they are planted with the latest improved varieties to multiply seed for commercial production, the key link from research to application. The sight of a solution to the wilt problem in these fields is enough to give one the sense of a case being closed, but this case is not closed. We leave the test field then and our taxi follows a motor scooter across the highway and down a farm lane to visit a farmer, a poor man, who is growing an improved variety. No wilt. Giri and Harsulkar hit the field and in a matter of seconds have scooped up a dozen or so pod borer larvae. There is no resistance to be had yet for the pod borer, but this poor farmer has at least bettered his odds by using wilt-resistant varieties. His yield is going to suffer some loss to the borers, but still he'll do all right, even without insecticide. He's farming reasonably well by hand, and the improved seed alone will bring him an acceptable harvest.

We head back to the highway and see another farmer out in his field. Most of his mud-brick house is caved in, the roof half off. A clutch of kids stands in the yard and stares, vacantly, as does the farmer, the nobody-is-home stares that look out from the world's most dehumanized level of poverty. The man has maybe a half acre of land, rocky and thin. He's planted his chickpeas too late, for no special reason. He's used whatever seed was cheap in the local market. He's weeded carefully and tended, but still his plants are pathetic, devastated by drought and wilt. What he needs most to keep his kids from starving is knowledge and improved seed. Both are literally within sight, just across the road, growing in an agricultural school's test plot.

Later Giri, Harsulkar, and I are having coffee at a rest stop, and they are talking about papers published as a result of their research, a conversation I interrupt with a rude question: What good are scientific papers to that farmer we have just seen and his malnourished, hollow-faced children? They take no offense. They are from villages, and this unsettling question is also theirs.

Milind Ratnaparke and Dipak Santra are in a sense both payoff and promise, payoff from India's long investment in the intellectual infrastructure. Already they have published the results of distinguished research, which they are reviewing for me at the National Chemical Laboratory in Pune.

Santra, the son of a village teacher, read science journals as a child. An earnest, impassioned fellow, he frequently lapses from the drone of scientific explanation into digressions about village life.

"I know what the farmers are facing," he says. "So I want to do something for the farmer, the poor farmer."

When he goes to visit his family, he leaves behind the world of biochemistry for a village where most of the farmers are illiterate to the point of not understanding the basics of fertility.

"They don't use green manure crops. They don't use dung. They use it for fuel. They are ignoring that we are killing our next generations," he says. "They don't know." That last phrase becomes a monologue.

"They don't know. And somehow we have got to fill up this gap. Whatever the existing cultivars are, with proper management you can still get good yield, but you've got to stop this leakage of information. You've got to start training. They just don't know. People have been appointed by the government to transfer the technology, but somehow that is not happening. I don't know why. It is their job to transfer technology from the lab. Somehow that is not happening. I know that the government has taken care of the system, but somehow it is not working."

It is not that there is a shortage of people in the villages offering information but, as in much of the world, that "information" comes from marketers, especially pesticide salesmen.

"I have seen in my village the pesticide companies. They are announcing . . . some vegetable . . . like cauliflower, like cabbage . . . you spray with this insecticide. This is what the marketing people are doing. That's the way it is going on at the village level, at the grass-root level. I am telling the fact, because I am from that. When I go, it just bothers me. It irritates me because I can't help them."

Dipak wonders what good his work is if it stays in his lab or on university test plots. He tells me he is searching for some way to break out of the mold, maybe give up the lab or take a month off every year and volunteer in his own village as a teacher of farmers.

It is easy to get lost in the science halls, wound up in the layers of sophisticated research that is producing a hypertechnology that can feed the world. This preoccupation with research ignores the fact that if all chickpea fields produced like Wane's, as they could in theory, India would not have a protein problem. Wane uses existing, readily available technology. Virtually all the world's agricultural research efforts are directed toward producing new technology, yet existing technology would feed a lot of people if someone would just figure out why it remains on the shelf.

# The Critical Mass

## The Fate of Farming in an Industrializing World

### [Nanjing, China]

The two most populous nations on earth are India and China. Together they include just over 2 billion people, a third of the world's population. This mass cannot be ignored, because it is sufficient to determine the world's future. India occupies a key position, now and historically, by virtue of its role as the vanguard of the Green Revolution. Where India finds itself, much of the world finds itself, in an almost paradoxical state of sufficiency, surfeited in carbohydrates but protein-starved. Its solution emphasizes simplicity, vegetable protein, subsistence, and rural self-sufficiency—a balancing of diet based on the realities of traditional culture and agricultural possibilities. This represents one model for a sustainable future. China defines the opposite path.

The 250-kilometer drive from Shanghai to Nanjing is surreal, made so mostly by the road itself, a brand-new freeway. China is not a freeway kind of place. Even in Shanghai, a shining city of 13.4 million, quiet neighborhoods exist because there are few cars. Instead of the roar of internal combustion, a green light releases a crescendoing whoosh, like a big wind building, the collective rush of a stream of bicycles. Even on the freeway one does not see many vehicles. The few cars visible are new, sleek black sedans China manufactures in a joint venture with a European automaker. They whisk along at 60 mph. Industry rules the landscape: a pall of foul air hangs unbroken even by rural sky. Everywhere we look we see industry, villages of sprawling factories, mostly new, with European names: more joint ventures. Every couple of kilometers another village crops up of three to four thousand people in a cluster of single-family homes, like townhouses in the United States, many of them new. Each village has its factory surrounded by square, meticulous fields of spring-green wheat that look like extensions of the factories.

We are in the lower Yangtze Valley, the most industrialized region of China and the most populous. Like the highway, it all looks new, but industry has been here for a long time. The lower Yangtze acts as the cradle of this civilization. Chinese people have been growing grain and building tools here since the Stone Age, in one of the longest unbroken stretches of human enterprise on the planet. It is the most clear-cut example of what the earth can become if humans are left in charge. In a 120-kilometer drive on this highway one spring day, I did not see a single bird.

It would be wrong to say that this represents all China, in that China is a hugely diverse place, with all the variations of custom and culture one would expect in a country that spans a continent and maybe five thousand years of recorded history. Still, the hyperindustrialization of the lower Yangtze Valley is one model of what China wishes to become.

We have come here to consider agriculture, but it cannot be considered in isolation from its context of industrialization, an engineered landscape, and, if the Chinese have their way, an engineered foundation of genes.

In 1995, a book titled *Who Will Feed China?*, by Worldwatch Institute president Lester Brown, focused concerns about the world's food security on that country (which made the subtitle, *Wake-up Call for a Small Planet*, a slightly odd choice). By the time I visited China four years later, the Chinese believed they had an answer: the Chinese will feed China. Indeed, the Chinese can grow grain. In the four years since, both the Chinese and others have criticized Brown's alarm as a Chicken Little report on food. Yet in many ways the ultimate outcome of Brown's forecast is not so much the point as his revealing a profound shift in the entire world's food supply as a result of the cultural evolution of China. China is industrializing, and its people are getting richer. Suddenly. Clearly industrialization and wealth were not unprecedented on earth, but with 20 percent of the world's population, China is an elephant in a rowboat full of small animals. All of the passengers save the elephant can move about without rocking the boat, but the ride remains smoother if the elephant makes no sudden moves.

Brown writes:

> Past experience has not prepared us well for assessing the scale of China's future food demand. Multiplying 1.2 billion times anything is a lot. Two more beers per person [per year] would take the entire Norwegian grain harvest. And if the Chinese were to consume seafood at the same rate as the Japanese do, China would need the annual world fish catch.

Aside from the sheer enormity of its population, a second factor is at work here. There is a huge inequity in the way the world consumes its grain. The average Indian, for instance, consumes about 200 kilograms per year, while the average American takes four times that much, because most of the American's consumption is indirect: grain fed to livestock for meat, milk, and eggs, and, with gasohol, grain fed to cars. Historically, as countries have industrialized, they have moved up the food chain from India's end toward the United States' pattern. China, too, has already begun that climb. Between 1975 and 1995, pork production increased sixfold. Per capita meat consumption has quadrupled since 1975.

This factor alone has been enough to push China's per capita annual grain consumption to 300 kilograms. China has responded to that demand with an impressive increase in grain production. Simply put, China is now the world's largest producer of all three major grains—wheat, maize, and rice. There is, however, no letup in the trend toward increased demand, which is projected to reach 400 kilos per person by 2030—a total, Brown says, of 641 million tons of grain a year. To meet that demand, China would need to import 369 million tons of grain annually, which is nearly double existing world exports.

We have no reason to think China's production can rise to meet that demand, and some reason to believe it will fall. While industrialization fuels the demand for grain, it also undermines the ability to produce it. Brown cites three other Asian countries as examples: Japan, South Korea, and Taiwan. All three have undergone rapid industrialization in recent decades. As a result, Japan lost 52 percent of its grain lands, South Korea 46 percent, and Taiwan 42 percent.

The lower Yangtze Valley is considered the cradle of China's civilization for a reason: it contains the country's richest agricultural lands. As a result, it also contains the bulk of the people, and thus the bulk of its sprawling new factories and highways. Richer people buy cars that need new highways, and China is building about 10,000 new miles of road per year. Richer people want bigger houses that take up more land, as do the factories. The pattern is worrisome, not so much because it is unprecedented, but because it is familiar. The Chinese understand all this. More than one visitor has complained about the pall that passes for air in most cities. The Chinese acknowledge that industrialization is a messy process, but also point out that we Westerners need only read our own histories for proof of this.

In a conversation with an American reporter in Shanghai, a long-time resident of that city, I ask him about the Chinese perceptions of the world's reactions to their treatment of Tibetans. He tells me he frequently asks Chinese officials about the situation and gets a similar response each time: What's the big deal? You Americans killed your Indians when you needed the natural resources they controlled.

Indeed, we Americans did.


While not Shanghai, not an emerald city of skyscrapers, Nanjing still rises. It has a spanking new tower of a Hilton hotel, flush with foreigners who are foreign nowhere, doing business. A bustling city center is ringed by glittering department stores stuffed with fashionable goods that would no doubt have shocked the dour Communist leadership of a generation ago. Still, there are back-alley mazes of open-air markets, noodle stands, herbalists, laundry, and freshly plucked chickens swinging in the breeze. Nanjing is a walled city, once the largest walled city. Visitors can still climb standing sections of the wall with gates massive as palaces and wonder what sort of bureaucracy it took to assemble the labor for such an enterprise six hundred years ago. At times the imperial capital, this city has both an ancient and a sad recent history. A new memorial near one of the city's gates tells, in wrenching detail, the story of the Rape of Nanjing, the occupation by the Japanese during World War II that brought a massacre of 300,000 civilians, many beheaded in sword-swinging competitions for Japanese soldiers. The Yangtze floated bodies.

Nanjing also has a first-class agricultural university. It lies just outside of the eastern edge of the city, but is hard to pick out on the horizon, even from the vantage of the section of wall that still forms the city's east gate. The university's blocks of high-rise apartments blend with the rank on rank and file on file of high-rise block apartments in lines that stretch in all directions until they fade into the smog.

Even dedicated to agricultural research as it is, the university plays a standout role in this particularly industrialized province of China. This shows in its orderliness, its new buildings of marble, newly stocked labs, and conference rooms with ornate hardwood furniture—all of which highlights the prominence of agriculture in Jiangsu Province, of which Nanjing is the chief city. It's small, containing only 1 percent of the total land area of China, yet it contains 6 percent of the total population, 71 million people in 102,600 square kilometers, almost 700 people per square kilometer. Its population density accrues from its long history as the center of China's civilization, which in turn was dictated by the broad and fertile Yangtze Valley. Still, it contains, in 1 percent of the country's total area, about 4 percent of the country's total croplands, but the disproportion extends further. It produces about 10 percent of the country's total crops.

Every year the nation holds a competition to identify the 500 most developed cities, meaning the most industrialized. Of the top ten, five are invariably in a single province, Jiangsu. Still, the province is overwhelmingly rural, 50 million rural residents of the 71 million total. The province feeds itself and the neighboring sprawl of Shanghai. All these statistics make Jiangsu a sort of indicator province, not so much typical of China as a distillation of China, an exaggeration of the seemingly contradictory trends driving the whole.

Liu Dajun directs the Institute of Cytogenetics at Nanjing University, the lab housed in the newest and best-stocked of the marble buildings on campus. A quiet, graying man, he has a kind, thoughtful voice, but the day we spoke, a bit racked by the flu. Still, he joined us on a foggy, rainy afternoon in a meeting. The Chinese tend not to heat public spaces, so very often rooms with their cold-holding masonry construction feel colder than the spring that breaks outside. On that rainy January afternoon, as we sat around the conference table in coats, we sipped warm tea to crack the cold.

"I would say," says Liu, "that the Chinese farmer is the best farmer in the world."

He backs his statement by quoting rising yield statistics as impressive as any in the world, a short history of the steep increase that paralleled the great leap of the Green Revolution.

"We will never give up production in the field," he says. "It is unreliable to rely on imported food. The government has stressed the importance of self-supply. We have to produce all the food ourselves."

He tells me that when the rising affluence of places like Shanghai created a lucrative market for vegetables, few farmers began to diversify to fill that niche, so the government provided incentives to get them back to the real business of cereal production. "Grain is most important," he says.

A few blocks away from the university stands the cavernous Nanjing Museum. Its extensive collection can begin to impress upon the visitor the depth of this culture and the patience with which it has honed its central ideas. On display is a quote from a book from the Han Dynasty, a two-thousand-year-old sentiment: "A king's life depends on the people, and the people's, on the food."

The name of Liu's lab, the Cytogenetics Institute, tips off where this story is headed, but it begins in wheat. If the surrounding fields are so central to China's agriculture, one would assume the topic would be rice, but wheat is only slightly less central, both now and throughout the course of China's agriculture. True enough, China is the leading producer in the world not only of rice but also of wheat. The Yangtze Valley, a broad horizontal strip, demarcates the two crops. To the south, rice dominates both agriculture and diet. To the north lies wheat country.

The cultivation of rice here is older than China's recorded history, with kernels, tools, and the culture of this agriculture showing in remnants that date to the neolithic. Wheat is an import that traveled from its center of origin in the Middle East along the Silk Road. It has been in the country for at least three thousand years.

Farmers in the less-populated and colder northern provinces grow wheat American-style: on large, mechanized farms as the primary crop. This practice has founded a regional diet centered on noodles and dumplings. In the hotter, more humid south, wheat takes on a secondary role to rice in both importance and time. During the summer, farmers plant their tiny, flat, and meticulously worked flooded fields to rice. During winter, these same plots produce wheat.

Moisture and heat are problems. Not just in China but worldwide, wheat grows in the temperate regions for a reason. It is a grass that evolved all its survival tools—its attentiveness to light, its rhythm of growing, its armaments against pests, and its architecture—for places with alternating seasons of warm and cold, not warm and wet seasons alternating with warmer and wetter. Pushing wheat beyond the edge of its range creates problems, the most pronounced of which is wheat scab.

Scab is a fungus, a Fusarium. It forms on the heads of maturing plants to raid the seeds of nutrients, typically in cycles that peak in alternate years or in every third year. A mild year sees yield losses of 10 to 20 percent as the Fusarium shrinks seeds. In a bad year yield loss might be 40 percent. Further, the fungus itself is mildly toxic to humans, so severely infected seeds can't be used for food. Any wheat showing more than two parts per million of the mycotoxin can cause vomiting in anyone who eats it and is thus useless for food.

The world's first case of scab was identified in the United States in 1891. Since then it has been discovered in southern Canada, Brazil, Argentina, Paraguay, Uruguay, Mexico, Japan, Poland, the Netherlands, the United Kingdom, the Czech Republic, Russia, and Austria. Still, plant pathologists believe it is endemic to central China, an important clue as to sources of resistance and enough reason to center a research program to combat it at Nanjing Agricultural University.

What is at stake here is more than the yield loss. Scab serves as a sort of limiting factor to wheat, a border on its range. Beating it back allows further advance of wheat into areas that are hotter and wetter, into rice regions, where it will not supplant rice so much as complement it like a second dry-season crop, as it now serves China. It moves the resource closer to the tropics, where most of the world's hungry people are, an irresistible advance to food producers, no matter how much it may be going against nature's grain.

So far, that advance has been made largely with the help of fungicides, still a primary tool of Chinese farmers. It is hard to imagine their widespread use as an improvement, even though a leading fungicide does have a name in Chinese that translates as "Let the wheat disease be peaceful." But as the Chinese researchers are explaining this to me, their voices drop to a serious near-whisper. They say that in the 1950s farmers controlled scab by spreading highly toxic mercury on their fields.

Fungicides, however, are not the only defense. Elsewhere in the world, especially in Europe, breeders have developed varieties of wheat with considerable resistance to scab. Their success led the Chinese to import and screen some of those varieties. Just as there are varieties of wheat, though, there are varieties of scab, and China's brand of pathogen was able to overcome the resistance in European varieties. That led to a breeding program in China with some screening of local cultivars to identify resistance. Eventually, a husband-wife breeder-pathologist team developed a resistant variety that remains the standard in much of Asia, but its resistance is only partial. The consensus was that the work would have to go beyond breeding. Enter the Institute of Cytogenetics.


Chen Peidu, Liu's colleague at the institute, is the other principal researcher in the wheat scab program. He grew up in Jiangsu Province, in the nearby city of Wuxi, and took both his master's and doctorate in plant genetics and breeding at Nanjing Agricultural University, an all-China education broken only by two brief stints at Kansas State University, in the middle of the U.S. wheat belt. He is a short, stout, straight-ahead kind of guy with a perpetual smile.

He smiles especially on a tour of his gleaming new lab on a rainy

Saturday afternoon, even then busy with grad students, post-docs, and lab assistants turning the nuts and bolts of molecular biology. Chen wants me to have a look through a microscope where a researcher has stained and prepared a slide. There they are, at 600 times the size of life: squished-flat chromosomes. It's an odd sensation looking at this straight on. Over months of discussion with molecular biologists, I had become used to thinking of DNA and the double helix and chromosomes as abstractions, like atoms, models that illustrate an idea. It's easy enough to forget that to many people, those with labs with 600-power microscopes, these are not abstractions but as real and accessible as the calendar hanging on the wall.

Much can be gained by simply looking. The Chinese have done what most plant breeders now do, which is to compare the chromosomes from resistant varieties with those from varieties without resistance. In this way the researchers located the site of resistance. In fact, there is not just one chromosome in play but three, meaning the plants have evolved a complex defense mechanism, just as the fungi have developed several varieties of attack.

These genes encode the present state of this plant, but they also show its history, which is relevant to the effort. Chinese breeders were unable to use the European resistance to scab, but did find local sources, as evolution would suggest. During two thousand years of cultivation in the Yangtze Valley, wheat interacted with local scab, and any tricks learned in the process are recorded in the chromosomes. There are, however, other relevant histories, a fact the Chinese know.

For instance, the scab itself is now believed to have originated in China, despite the fact that its host originated in the Middle East. This means scab infected other hosts long before wheat arrived on the scene. The fungus is something of a generalist, in fact, plaguing cereals like rice and rye, but also the larger family of grasses in general. It is not, however, a serious threat to any of these in that the relationship is sufficiently long-standing to allow the older plants to evolve immunities. Scab gets by, simply survives, on the others, but it thrives on the newcomer wheat. Researchers could exploit this simple fact.

While they continued with simple breeding techniques to try to beef up immunities within the species, researchers also escalated matters by working in the next concentric circle out, by crossbreeding with wild relatives by hybridization. A search turned up three species of wild perennial grasses with very high resistance to scab, one growing on Nanjing's campus. Researchers set to work attempting to crossbreed the wild grasses with wheat to transfer the genes for resistance, but because they had already established that the genetic basis of that resistance was a complex—meaning three or more chromosomes were involved—transfer through breeding would be a messy, hit-and-miss process. Further, there was not a natural pairing of chromosomes between the species; the respective genomes simply didn't match up. Crossbreeding essentially went nowhere, even with complicated tissue-culture techniques.

That left only one option, the step into the next concentric circle: fairly simple, straightforward genetic engineering. Using three different established methods and the knowledge they'd gained by genetic mapping, the researchers successfully transferred resistant genes from the wild grasses to wheat. The genetically engineered varieties provided the material for further crosses, backcrosses, and screening—in other words, fed into a conventional breeding program to select for the various agronomic traits necessary to introduce broadly resistant strains of wheat.

I had asked Chen to show me not just his lab work but something about the situation and practices of farmers in his region. This posed a bit of a problem in that unlike most of the other McKnight programs, the researchers at Nanjing do not work directly with farmers. But it was arranged and fit into an itinerary that resembled a formal state visit more than a journalist's research: sightseeing, full-dress meetings with dignitaries, and nightly formal banquets.

The banquets allowed plenty of time for me to memorize Chinese cuisine, especially as the evenings wore on, beginning with polite toasts clearly translated to be relevant for the visitor but lapsing to nonstop, back-to-back rounds of toasts, untranslated, but mostly relevant to the various personal relationships around the table of ten or so. Meanwhile, at the center of the table an endless array of favored foods turned on a lazy Susan, a fixture of banquets: pork, chicken, beef, shellfish, squid, jellyfish, deep-fried sparrows. Not everyday fare, but the face one shows guests.

Then back on the freeway, whizzing along in a black sedan at 120 kilometers an hour. Chen wished to show me the farming area around Wuxi for much the same reason as there was meat on every table. Aside from the fact that it is his hometown, it is a paragon of sorts in the Chinese model of development. Along with two adjacent cities, it usually places annually in the top-ten list of the nation's most-developed cities.

As we sped through Wuxi, I noticed that our entourage was growing, as cars carrying various agricultural officials from the region joined the procession. All this lent a certain air of authority to our arrival in Huang Ni Ba village, as drivers whisked up in front of the village's central office buildings and sundry greeters began opening doors. En masse, we, now some fifteen—researchers, bureaucrats, translator, aides-de-camp—snaked around the perimeter of a large meeting room ringed with couches and chairs. The room was unheated, but tea was served and duly slurped. The head of the village arrived, carrying, as is customary for the headman at an important meeting, cartons of cigarettes. These he opened with a flourish, and tossed a pack to each person present. Then, with another flourish, he opened one pack and began to distribute a couple of cigarettes to each person present, a gift to save them from having to squander the full-pack gifts during the meeting. Squandering began immediately all around, and as the last flick of the Bic dimmed to an afterglow, I stared out at a full circle of fifteen earnest faces, each intently peering back at me over poised cigarette and through the gathering haze. The translator provided by the university, the indefatigable Miss Han, herself a nonsmoker, coughed a couple of times, then said the interview could begin.

I asked a question about the state of farming in the village. The village headman launched into a long discourse consisting mostly of regional agricultural statistics: X tons per hectare of rice this year, X tons last, the same for wheat, a history of success. It was the same list Chen had already recited for me and the same list I would hear again and again in response to almost any question about farming. To the Chinese, farming is yield. I tried a follow-up, but somehow that got warped into a question about converting kilograms per *mu* to tons per hectare. The room erupted into a full five minutes of debate on a simple calculation.

I tried another tack, straight on. On the way to the meeting, the motorcade had glided through a series of fields that surrounded the village, a set of perfect, pool table–flat squares bordered by concrete ditches and paved concrete roads, each and every field fallow after the harvest of wheat, awaiting rice. There were vegetables growing, too, but not in the fields, instead along ditch banks and in back yards. The land assigned to agriculture, a total of 110 hectares for the village's 540 families, was for fields of grain as regular as graph paper.

I asked about this monoculture, whether there were plans to diversify or for crop rotations that included legumes or vegetables. This issue has a rather direct bearing on the scab problem, in that the monoculture of cereal has created an almost perfect environment for propagating scab. Even when wheat is absent in the off-season, the disease moves into rice to survive until the next crop of wheat.

"No"—a general answer from the haze—the farmers have no interest in growing anything but rice and wheat. And yes, monoculture is a problem, but the farmers believe that's what guys like Chen are for. Researchers and their powerful technology will solve the problems monoculture creates.

To a certain degree, the villagers are free to lever changes. The village head controls the cigarettes because he has real power in this system. After the latest round of political liberalization, the control of the land and the economy of each village passed to the people of the village, who elect their leaders. Each village family has its own plot of land, and the village also maintains communal plots, but the concrete ditches, roads, fertilizer, and applications of "Let the wheat disease be peaceful" are paid for by the government, and repaid in an in-kind tax of sorts, in that each village and each farmer owe the government a certain amount of produce each year. The government wants that tithe as cereal. Once the tax is paid, the families are free to sell what remains, but the system is set up to buy cereal.

"With wheat and rice we can get a good yield, and the farmers prefer it," the village headman told me.

They prefer it because it is the only way for the surrounding landscape to support 700 people per square kilometer. Cereal is at the center of world agriculture because it is essentially a dense pack of carbohydrates, a food factory. It creates the possibility of civilization and high population density, but dependence on it guarantees a future of more cereal, which leaves people like Chen no choice but to navigate the dense information of the double helix.

We leave the meeting to visit some village houses. All alike, they stand in a cluster, brand-new white-glazed brick townhouses, each a single-family home. Each has a full complement of major appliances: wall heaters, washing machines, and gas stoves. We see comfortable living rooms with carpets, some marble floors, brightly colored tile roofs, windowed alcoves, tiny, neatly landscaped yards. Families tend to be extended, usually having about four wage earners per household. Each family owns its own house, for which it pays about two years' worth of wages from four people—wages that are about double the average for the region. "They are rich," Miss Han tells me.

The cause of their prosperity has little to do with farming. Just as they are free in theory to decide what and how to farm, the villagers are likewise able to enter other enterprises. This village has established a large machine shop for heavy manufacture of gas and aluminum pipe. Most villagers work in the factory, the source of the handsome wage.

Still, each village is land-based, so the houses cluster at the center of the 110 hectares of farmland, which must be farmed. Each family also has some responsibility there, but increasingly it is the practice of these prosperous factory workers to hire people to farm. Farming is an afterthought, a necessary bit of business on the way to prosperity. Mostly the result of recent reforms, such changes have altered village life throughout China, but especially in this central region. Miss Han, a young professional woman of the city, bright, well-educated, with smooth English, grew up in a village that also has developed local industries.

"My mother worked in the fields day after day. Now she is free," she says. "Township enterprises developed, and now the villagers hire people to grow wheat and rice, and they pay them."

Meanwhile, a generation later, for her own young daughter, the lure has changed to the burgeoning string of McDonald's franchises in Nanjing. Miss Han has finally bargained her way down to a trip to McDonald's no more than once a month.

Chen has arranged for me to see the Wu Culture Park near Wuxi, a sprawling park-museum that is normally closed the day we visit, but officials open it for us. In the various buildings of the complex are displayed tools and machinery, mortars, pestles, water wheels, grinding wheels, artifacts of seven thousand years of rice-growing history. Multiply this depth by the breadth of China, by its sheer mass of population, then remember that we are visitors here. We may watch, but anything we think, say, or do will have no bearing on this mass. China will do what China will do.

Just before I left Nanjing, Miss Han and I talked about a downside to the reforms, the increasing unemployment brought on by dislocation of some old, subsidized sectors of the economy. She said the Chinese used to have a saying that covered secure employment: "I have an iron rice bowl." The Chinese no longer take the iron bowl for granted.

# Genetic Revolution

## Bioengineering on the Loose

## [Shanghai, China]

Su De-Ming is explaining genetic markers in a room full of trained geneticists.

"Su is the one with a bald head and glasses," he says. "So if you had all the bits of everyone, you would not need to see all of Su. Just bald hair and glasses to know Su was here."

Close enough for genetic markers, but to pinpoint Su, one would also need markers for ebullience, gregariousness, and a quick sense of humor, traits that would make him stand out in a room full of scientists almost anywhere. Su runs the Virology Research Unit at Fudan University, one of the nation's oldest and, by virtue of its ties to the growth engine that is Shanghai, one of its leading universities. Su waxes most ebullient when he is talking about Shanghai, as he does one sunny afternoon when we have a full view of its skyline of 4,000 new skyscrapers. We're getting a rushed view, part of Su's point, as we whiz across town in a van along the city's new freeway. He tells me this drive was not nearly so fast only a few months ago, what with growth, congestion, and traffic. Then there were only two lanes, but a third lane has been added, so the future appears brighter.

"It is best to have three lanes," Su says.

And Shanghai aims to have the best, having long chafed under

Beijing's control, despite its own long history as an international center of trade. For instance, a few years ago the city built a brand-new soccer stadium, only to have the central government limit its size. "We were not allowed one seat more than Beijing's," Su notes.

But that sort of pettiness is fading, especially, he confides, now that Jiang Zemin, a native of Shanghai, is President. "Shanghai is now the paradise of China. We are lucky. We are a lucky people," Su says. "Now you can have everything. Now it is so easy."

In keeping with Shanghai's spirit, the project that won a McKnight grant at Fudan is quite bold. The researchers mean to blaze a new path in fighting not just plant diseases but disease in general, a plan made evident by the fact that the partner institution in the United States is not a land-grant school, as is the case with all the other projects, but the Yale University School of Medicine. Su is joined in this endeavor by Shen Daleng and Li Chang-Ben, the project's principal investigators. All three are middle-aged men, which makes them an exception. The rest of the staff is much younger, and there is a clearly evident age gap. The absence of a generation, this dark spot, is really the shadow of the Cultural Revolution, an event that shapes the thinking of the project as surely as Shanghai's skyscrapers and technological optimism.

Not a lot of farming can be seen around Shanghai; those new skyscrapers have swelled the city, spawning suburbs where there used to be rice fields. The small agricultural institute nearby where the project does its fieldwork has a hard time holding its own against the high-rise apartment buildings that walk right up to the fields' edge. There are no farms around, so the test fields now sample not a farm environment, with its typical load of pests and plant pathogens, but Shanghai's air, roughly the equivalent of two packs a day. Lucky for the work, this project does not have a lot to do with what happens on the farm, although it centers on rice, the column of carbohydrates that supports the weight of Asia's population.

Fudan's project takes aim at rice stripe virus, a disease probably as old as rice itself here in the lower Yangtze Valley. On average, the virus takes a toll of about 5 to 10 percent of the total rice harvest each year. This percentage is not particularly staggering, especially when measured against the losses to disease in crops such as chickpeas or sweet potatoes, which are near total. Ten percent of the rice in Asia, however, is a very large number.

The Chinese fight rice virus primarily with insecticides, so the project is aimed at lowering pesticide use as much as improving yield. Insecticides work because the virus has an alternate host to rice, a miniature grasshopperlike creature called the medium brown rice plant hopper, *Laodelphax striatellus*. On its own, the barely visible hopper does very little damage to rice with the bits it sucks from stems, but in doing so, it also injects the virus. The fates of the three—virus, hopper, and rice—are linked. This sort of linkage of life-forms underpins the complexity of the natural world. It is a hallmark of the emerging, postindustrial design of technologies that they don't attempt to paper over or ignore this complexity but instead seek ways to exploit it. If complexity is not a barrier but a tool, then it would seem the more complexity the better. There can always be more. Very few ecological relationships that mean anything at all can be fully characterized by examining only three players. This one is no exception. The plant hopper itself is barely visible, but when the scale of vision, the order of magnification, is expanded, more players emerge.

Battling viruses is a tricky business that requires antibodies, but plants don't have immune systems, so they can't make or even be induced to make their own antibodies. As an alternative, it would be a relatively easy matter to develop a vaccine against rice stripe virus. Viruses work by entering the cell of a host and, in effect, taking over the controls to produce more virus instead of the cell's normal product. The relationship is intricate, but what allows the virus to splice its way into the cell is its geometry, which is to say, it can go only where it fits. Animals make antibodies, special proteins that have a specific configuration that allows them to splice into cells to prevent the virus from doing so. Scientists can make antibodies experimentally in animals, even an antibody to the rice stripe virus. They know the shape of the rice stripe virus and so could design a mirror image to defeat it. That's the fundamental principle of making a vaccine— but what good is a vaccine when it isn't practical to vaccinate every rice plant in a field?

A second problem with attacking the virus in the rice concerns food. Attacking the virus in the rice means launching an amino-acid arms race inside something that appears routinely on tables in most of the world. There's no telling where such a fancy might lead, but it will clearly affect the human digestive system, which has its own issues with antibodies and amino acids.

In this case, though, the hopper, which is an animal, does have an immune system to tinker with. Indeed, the hopper probably would have developed some immunity by now if evolution required it, but the virus seems to be along only for the ride. There seem to be no natural immunities of the sort that can be teased out and exploited by breeding. Perhaps a way can be found to give the hopper a gene to express a protein that will defeat the virus, but that seems too bold a path as well. Transgenic plants and animals are now a part of our world, but they are only a part of the artificial world so far. That is, there are relatively few individuals, and they are invariably domestic and thus within our control. To create a transgenic insect immune to the virus it normally hosts is to begin the process of altering the general wild population at large. Even if the technical means were available to accomplish such a thing, no one, not even the technologically optimistic Chinese, is ready to try an experiment on that scale.

But a smaller scale might work.

The trained naked eye can find plant hoppers on rice, but it isn't at all easy. They're that small. Still, a post-doc shows me a row of the insects pinned by air pressure to what amounts to a tiny examining table beneath a microscope. The bugs are face up, and the re-searcher is guiding a superfine needle toward the soft joint between hard sections of carapace. She is giving each of the hoppers an injec-tion, the trick being that they must survive, not for themselves, but for yet another organism they host. The target of this process, called "microinjection," is even smaller than the hopper. The real star char-acter of this story is a weird little microorganism called *Wolbachia*.

*Wolbachia* is a symbiont, a being that can jar our notion of what defines an individual and a species. Symbionts are single-celled crea-tures, bacteria, that live exclusively inside another creature. The host's and symbiont's life cycles are so entwined that the microorgan-ism simply cannot survive outside the host. Still, it is a separate crea-ture by virtue of having its own DNA and its own distinct evolutionary history that was once separate from the host but is now inextricably linked. A form of *Wolbachia*, for instance, lives in the African tsetse fly, and evolutionary studies suggest it has done so ex-clusively for a million years.

The border between host and symbiont, however, is not nearly so distinct as it might seem. Most plant cells contain structures called chloroplasts, the basis for photosynthesis. Most cells, including those of humans, contain structures called mitochondria that help cell mo-bility, another vital service. Clearly, these mitochondria are part of our being, yet DNA studies confirm suspicions that these structures have a separate evolutionary history from the rest of the cell. They were once separate, like symbionts. What, then, is an individual?

In the case of the plant hopper, the boundary between host and symbiont is still visible. *Wolbachia* may not be capable of indepen-dent survival, but it has its own DNA. Furthermore, it is a prokary-ote, a primitive sort of cell without a nucleus, which makes its genome more accessible. So if not the rice and not the hopper, how about *Wolbachia*? Can it be taught the trick of creating antibody pro-teins and become a microscopic and ubiquitous factory producing rice stripe vaccine?

Getting an antibody is the easy part. The path there is well worn through experiments with rabbits and mice. It is a matter of placing the virus in an animal and letting the mammalian immune system go to work. The researchers did that and came up with a list of proteins produced to counter the virus. They narrowed the list according to what they had learned about the virus itself. It does its damage by replicating and moving from cell to cell to infect large areas of rice tissue. It moves through small openings in cell walls, but, in fact, the virus is larger than those openings. It squeezes through by producing a movement protein that causes the opening to enlarge. Without the movement protein, the virus gets trapped and harmlessly imprisoned in a single cell. The researchers isolated two movement proteins from the virus, placed them in mice, and got antibodies to those specific proteins.

This step is crucial for more than just creation of a vaccine, in that an antibody works according to its shape, the configuration of the kind of protein it is. That structure, then, almost literally becomes the key. It's like finding a piece of a jigsaw puzzle. Finding the interlocking piece means you have found the gene that makes the protein.

Dr. Li, one of the principal investigators on the project, offers an analogy for the importance of this tool: "It's like a pool with a lot of fish inside. Now we have a pole that gets the most delicious fish."

Instead of giving the *Wolbachia* a fish—the antibody—researchers will teach it how to fish by giving it the gene for making the antibody. Once they can locate genes for expressing the antibody, it is a relatively straightforward matter of tinkering with the accessible *Wolbachia* genome, of genetically engineering it to express the antibody. The researchers are not engineering the plant hopper itself but rather a microbe that cannot survive outside the hopper. It is being taught to counter a plant virus through a precise trick. Chances are exceedingly slim that the tinkering will have any effects on animals who eat the rice, so the antibody is isolated in a different respect. The researchers at Fudan University are creating an extremely isolated process.

As it turns out, perhaps too isolated. Once they identified *Wol-*

*bachia* as the potential agent of this scheme, the researchers learned that the microorganism was so far known to inhabit only the gonads of the plant hopper. To do any good whatever against the virus, the antibody would need to be present throughout the hopper, in contact with the virus wherever it showed itself. The researchers segmented hopper bodies, then smashed up the DNA in each segment. That allowed them to sort through the bits—the equivalent of looking for Dr. Su by looking for glasses and a bald head. Eventually, in parts of the body other than the gonads, they found the genetic fingerprints of *Wolbachia*. It lived throughout the hopper, after all, one of the luckier breaks of the whole project.

The biggest technical hurdle lay ahead, though. It was an easy matter in theory to genetically engineer hoppers to express the antibody; the hard part was getting the whole natural population to do so. How much harder, then, to alter the whole population of invisible microbes within the whole population of plant hoppers? This is the question that leads the linchpin of the whole project, a phenomenon called cytoplasmic incompatibility. This is where the Chinese are charting new ground.

It was already known that *Wolbachia* infected the gonads of the insects, because infection affects reproduction by the symbionts on the host. Specifically, when an infected and uninfected hopper mate, only a fraction as many of the progeny survive as when two infected hoppers mate. An incompatibility favors unions of hoppers infected with the same strain of symbiont, a reproductive advantage for infection, and a trick the *Wolbachia* itself has engineered, no doubt to ensure its survival. The Chinese plan to harness that effect to let the transgenic microorganisms spread throughout the population of hoppers in China, and beyond, for that matter. No one knows whether it will work or how it will work or, for that matter, how one might stop this experiment if something goes wrong.

I walk a test field with Su where the researchers are doing some preliminary screening and testing in virus-infected rice. Most of the hoppers are enclosed in tight screenhouses, not transgenic hoppers, but virus-infected hoppers. The researchers are plotting the relationships that develop in tightly controlled conditions, but the real experiment here will be carried out in an uncontrolled environment. That's partly why Su can't answer my questions about how rapidly the immunity to the virus might spread. It depends on a host of imponderables like wind, weather, and the fortunes of the seasons.

"We don't have any dispersal data," he says.

Might something go wrong? He believes it's highly unlikely, given the precise nature of the tinkering, a narrow bit of change in a broad gene pool.

"I don't think they can cause any damage, but of course you never know how they might mutate."

But then that is true whether the hoppers are engineered or not. A mutation has probably occurred in your own body as you have been considering these last few thoughts, such is the fluidity of the genome.

Meanwhile, more hangs in the balance here than some stripes on rice. This little dance with virus, insect, and microbe inside insect is a common theme in the choreography of nature as well. Tsetse flies use this method to spread sleeping sickness. A similar path spreads malaria. Enough may well be at stake to merit a bold experiment.

It was really a matter of luck that I got the chance to buttonhole Frank Richards at a conference. He is the sort of source a journalist will spend a lot of effort to approach, simply because of his great knowledge and stature. Seventy-one years old when I met him and facing serious health problems, he is still a key part of the Yale Medical School, director of the Tropical Diseases Research Unit. He happened to be sitting next to me in a panel discussion, a short, stout man in a sweatshirt and khakis with the stereotypical countenance of the academic Brit he is. I began a cautious approach by addressing him as "Dr. Richards." He cut me off in mid-sentence: "It's Frank."

Richards provides the link between the Shanghai project and its partner, Yale University's medical school. Agricultural research in China pairs well with medical research for the same reason Richards himself works on shrimp: a subset of genetic engineering call paratransgenesis, the use of genetically engineered organisms to "passively immunize" against disease. The medical school is involved because ultimately human disease is targeted. The entire field takes advantage of the fact that an individual organism is not individual, that integral to each being's makeup is a vast collection of symbiotic microbes. Paratransgenics has a history as old as genetic engineering in that some of the original work in the latter field was done with *E. coli* bacteria, which are ubiquitous in animal digestion, including that of humans.

Paratransgenics has already produced a passive sort of immunization against Chagas' disease, or *Trypanosomiasis*, which plagues people in Central and South America. About 7 million people are infected, and most don't know it. It can lie dormant for ten to thirty years, then manifest itself as heart disease at a time when people expect to get heart disease.

"It's a disease that sits almost in the background of people's lives," says Richards. "It is carried by a set of bugs."

Past efforts have attempted to control the disease by attacking the bugs—*Trypanosoma cruzi*—with insecticides, a solution only partly successful and fraught with all the usual problems, such as insecticide resistance. Researchers sought another path and found that the insects carried a set of symbionts. Most do. *T. cruzi* is a bloodsucking insect, but cannot get everything it needs from blood and so relies on a string of microbes to digest its food, providing further nutrition, just as in humans. The researchers successfully inserted a gene in these symbionts that expresses an antibody to the disease, preventing them from spreading it to humans. The problem again becomes how to spread the gene through the general population of insects.

The answer lies in the way the symbionts are spread. It turns out the insects are born without the symbionts but have evolved a way of taking on the necessary load of microbes: by eating the feces of adults.

"So we made fake feces," says Richards, and throw in a healthy serving of genetically modified symbionts. "It sounds disgusting, but this is our way of spreading it. It isn't terribly popular to spray bug feces around your house, so we renamed it 'Cruziguard.' It seems to work. It's going into large-scale trials."

Richards calls this sort of mechanism, which exploits some aspect of the host's biology to spread a bit of artifice by natural means, an "engine." His metaphor seems to include the means by which science's advances are adopted in the larger political and economic system. Richards himself has learned to push a piece of science along until it attracts the attention of someone who can use it, someone with a big stake and big resources to pull it to the next level. The work with Chagas' disease has been "coopted," he says, by the U.S. Centers for Disease Control, which has now invested millions of dollars in bringing it to practical application.

Richards himself works in shrimp diseases, simply, he says, because they are economically important. Farmed shrimp have become a multibillion-dollar industry worldwide but suffer catastrophic losses from viral diseases. He is working to genetically modify the algae the shrimp eat in order to immunize them against these diseases. The shrimp-farming industry provides the engine that allows him to advance the basic science, both by paying for the research and allowing him to do his experiments in Asia, where he encounters fewer restrictions. Paratransgenics are becoming the path of least political resistance. That is, people may object to eating genetically modified shrimp but are less likely to object to eating shrimp that have eaten genetically modified algae. That's one way to isolate the controversy. A second is field-testing the basic science on agricultural applications such as shrimp and rice in order to work out the safety issues before moving into applications for human health. Even better still is to do that work in rice fields in a place as technologically optimistic as China.

"People get afraid of them with good reason," Richards says of genetically modified organisms, but he still doesn't believe that the possible dangers outweigh the benefits, so his strategy to skirt the controversy. "In my experience, it's much easier to climb a hill by the indirect route."

Somewhere up that hill, researchers will assault sleeping sickness. There might perhaps even be a path to widespread vaccination against a host of ailments through engineering of our own symbionts, such as the common lactobacillus eaten every day in yogurt.

A young post-doc, Chen Xiaoai, is working in her lab at Fudan while she chats about her role as a young scientist in Chinese society. She was the first to point out to me that only young and old scientists work here, that the gap of middle-aged people suggests a war or a plague of some sort.

She was in grade school in the early seventies, when the country was still relatively quiet—lucky timing, she thinks, in that primary education was still solid in China.

"I didn't know too much about science then," she says. "But one thing that attracted me was the human organism. It is like a miracle. I decided I wanted to solve this mystery."

The trail of her investigations eventually took her abroad; she did her doctoral work at Yale. This has been a typical path for many in her generation, and for many it has ended in the United States. Once trained, these young Chinese scientists stay abroad, overwhelmingly, Chen says, out of a belief that one can do better science in the United States.

Historically, there have been a number of solid reasons for this preference, mostly having to do with facilities and the opportunity to immerse oneself in a critical mass of scientists, to experience the stimulation of a true scientific community. Increasingly, though, such distinctions are disappearing. First, China has made substantial investments in facilities. Chen says there is nothing she could do in a lab in Yale that she cannot do at Fudan. Furthermore, there is a conjunction of two information revolutions.

The term "revolution" conjures an image of computers and microchips, but DNA codes information in much the same way that software codes instruct computer hardware, although the complex technology of the microchip has gotten all the press. People working as part of the parallel revolution occurring in mapping the genome believe that biotechnology is every bit as earth-shaking. Moreover, much of the former revolution fuels the latter; our insight into the biological code has been greatly enhanced by the computer's unmatched ability to sort and examine that code. DNA computers—computers that actually use living DNA as their "hardware," their circuit boards—are a state-of-the-art example, but in the day-to-day world of Chen and her fellow researchers, the tools of revolution are simply work stations with access to the Internet, through which they are connected to similar labs at Yale, at the Scripps Institute, in the United Kingdom and Australia. Chen tells me she feels if she is working in the same lab with her colleagues at Yale; the Internet has broken the national borders of science. The genetic code is being reconstructed in microchips in a fashion even more vital than an analogy to its existence in cells. It begins to live a new sort of life there. Chen left a China where scientists could not do good work and within a decade returned to one that works on the leading edge of biotechnology.

Nonetheless, Chen says she still believes it is better to do science in the United States.

While I was in Shanghai in the winter of 1999, a Chinese court sentenced a man to two years in prison for E-mailing a list of 30,000 Chinese contacts to an electronic publication based in the United States. The ruling was widely viewed as a warning to those making increasing use of the Internet, a warning that China can and will crack down. As is the case almost everywhere, use of the Internet in

China is so widespread and fluid that the warning is perhaps more accurately viewed as a hollow threat. People I spoke with there (although not people at Fudan) say they get what they want when they want it on their terminals. Still, the government remains inclined toward control.

The reason it's better to do science in the United States is independence, Chen says. But science is not all there is to her life.

"When I left this country, I missed it very much," she remembers. So do most of her colleagues who have not yet returned to China to do their work.

"Trained people go away. Almost all go to the United States. Most stay," she says. "After ten years or so, many will come back. That is my guess."

In a completely ordinary-looking building on Fudan's campus, we—a couple of Chinese researchers and I—climb into a cavernous elevator for a trip to the top floor. The anomaly in the scene, the elevator operator, a grizzled old man of sound proletariat credentials, sits on a stool in one corner slurping tea from a glass and pushing the buttons we ask for. The researchers whisper something to him, and he whispers back. They tell me he has advised them that the old man is in the building, that he came in very early, as he does every day. The elevator operator says this as a matter of pride that he is privileged to keep track of the venerated one's coming and goings. The researchers have asked because we are en route to keep an appointment with this old man, Dr. C. C. Tan. He's ninety-one, but most days he still comes to the School of Life Sciences, where he has been the scion since 1952.

Frank Richards has kept track of Tan over the years, with good reason. He said at one point he took a look around Yale and found thirty-five faculty and grad students who were Chinese geneticists trained in China. Richards says this is Tan's work. Chinese penetration of the field exists in sharp contrast to the Russian; there are almost no Russian geneticists of international stature. In the late 1940s, the leading Russian geneticist, T. D. Lysenko, decided that the Western view of the field didn't pass muster with socialist realism and set about revising the matter. He developed a picture of genetics based on inheritance of acquired characteristics, a picture he tried to promulgate with the Chinese, over whom the Russians at that time had huge influence. Tan, meanwhile, disagreed. He got the ear of Chairman Mao and was able to prevent the Russian version from penetrating. Put another way, as Richards does, Tan single-handedly saved the science in China.

Tan's office is small and crowded, but it has a comfortable couch. He wants to talk about the grand sweep of things. Li and Shen join us, showing great deference to the old man.

There are, he tells me, two key theoretical developments that define the twentieth century. One is Einstein's theory of relativity, the other is Thomas Hunt Morgan's mapping of genes. If he had to narrow that field, he makes it clear he would go with genes. He has some bias; Tan was a student of Morgan's at Cal Tech fifty years ago. Tan returned to China to found the life-sciences department at Fudan shortly after the Communist Revolution. He is considered the founder of genetics in China. This has not been a purely academic pursuit; scientists through this period had to struggle constantly for government acceptance, and the struggle continues. Even now, Dr. Li also acts as a citizen politician, sitting on the People's Congress for Shanghai. His colleagues consider this essential conduct for maintaining a good relationship with the mayor.

For Tan, much of this situation has amounted to a struggle to convince a country sinking all its resources into industrialization that the life sciences are essential, that the research he and his colleagues are pursuing will pay off. There was not a lot of payoff during the first twenty-five years of the School of Life Sciences.

"I call it," says Tan, "a period of trial and error. In China the biological sciences started in 1978. China has spent the last twenty years just catching up."

A new entity on Fudan's campus—a landmark of that catching up—will be called the Morgan-Tan Institute. It is meant to be a Taj Mahal of laboratories. Its primary function will be to serve as an inducement, a lure to be flashed before post-docs after they train abroad. In 1999, the institute was still an artist's drawing of a futuristic-looking building, but Dr. Li, who has already been named director of the new institute, reports having raised 5 million yuan toward its construction. Total cost is 16 million (about $2 million U.S.), but the government has promised 8 million if Fudan can raise the other half, so Li is optimistic.

The institute has another aspect. The core will be the lab at Fudan, but Li and the others plan a sort of satellite at Pudong, the epicenter of Shanghai's business boom. These researchers want a presence there in order to attract partnerships between the genetic engineers and international business.

The ebullient Su is officially Tan's successor and thinks often about assembling the brain power to drive his department's—and his country's—future. These questions are not abstract for him. He was educated in Canada and makes frequent trips to the United States for conferences. He uses those trips to test reality.

"In 1978, our instruments and facilities were out-of-date. So in the early eighties we got a loan from the World Bank. We imported many sophisticated instruments. It updated very fast at that time. We also sent associate professors abroad for training, especially in the early eighties. They were supposed to stay there one and a half to two years, but they stayed longer. There's another gap now, because the brain drain is really evident."

Then he ticks off a list of his students and where they are working now: Northwestern, Utah, Texas A&M.

"One of the problems is that even if they return, there are no good jobs. What do I mean by a 'good job'? High salary, but also good facilities. In this aspect we are still a developing country," Su says.

In 1993, Su was in the Eastern United States for a conference and made a list of forty Chinese colleagues all working in the vicinity of New York City. He set out to visit all forty of them.

"My intention was to invite some of them back, but after I found out what they were doing, I said, for the moment, better for them not to return. I asked them not to return. Because there is no possibility of jobs. No policy. No real policy for them to stay here. But now this is changing. There is a better economy, but problems remain. China is still developing. In Shanghai there are more possibilities, but even we can't match conditions like they have in the U.S."

Su delivered this whole monologue while we were riding in the van through Shanghai's new prosperity, in which he takes so much delight. The Chinese are true believers in progress and technology in much the same way that 1950s Americans were. American attitude was shaped in large part by the Great Depression and World War II; middle-aged Chinese like Su have seen their troubles, too.

"My own personal experience really synchronizes with the open-door policy," Su tells me. And synchronizes with the darker two periods that were predecessor to the current round of economic liberalization.

Su and Tan both weathered the Great Leap Forward, a period when Mao decided self-denial could bootstrap the social capital necessary for the country's progress. The Great Leap brought a famine from 1959 to 1961 that killed, historians now estimate, 30 million people, a toll beyond the imagination of anyone not used to thinking on China's scale. Shortly thereafter came the Cultural Revolution.

The way Su tells it, the years 1966–68 were the hardest. He and Tan were among those still on campus before being sent to the countryside for forced labor. The campus had been taken over by revolutionaries, workers who believed the professors' intellectual work was nonsense. There was no teaching, only re-education, which consisted mostly of harangues from workers interspersed with daily hour-long opportunities to study the works of Chairman Mao. Intellectuals were a particular target, because they had accumulated some status.

Tan's wife, for example, was forced to kneel in front of a wash-basin as the revolutionaries cut off all her hair. A few hours later, someone pulled Tan away from his office to tell him to go home. His wife was dead, a suicide. Su says there were thirty-three suicides among his colleagues at Fudan during those two years.

The remaining faculty was rounded up and sent to the country-side, where they worked on farms in 1969 and 1970. The intellectu-als were no good at farm work, and the farmers told them that they were a burden. There was not enough food, and a daily ration of rice—"no meat," Su stresses—left them malnourished. In the second year, the teachers got some students and some primitive labs. Su did some plant-breeding experiments and taught plant breeding and plant protection to whatever students he could find. He wound up writing a government brochure on plant protection. There's irony in his smile as he acknowledges that the government did indeed do some re-education; things he learned from the farmers and their lives still stick with him and guide his practices.

But what carried him, too, was English. He had learned to read English as a child because his parents were teachers and had access to some scientific journals in English. During the Cultural Revolu-tion, when reading English was illegal, he read secretly from pirated editions of journals. "I did it only during the night," he says.

Just as important as keeping abreast of the science was the lan-guage. When his country opened up again, he wanted to be pre-pared.

As he tells me this, the new highway arches to pass over a central city street below. It is awash with sleek new cars and sleek new peo-ple, dashing through hotel lobbies, stepping into cabs, flipping open cell phones and laptops, doing business.

"You see, we have the bright days now, but . . ." Su, who never seemed taxed for words in any language, never finished that sen-tence.

# Forging a Magic Bullet

## Technology Based in Biodiversity

### [Chile and Brazil]

A picture frames this story, a photograph I shot near La Serena, Chile. It shows a mound of potatoes in a gymnasium-sized shed full of seed potatoes. Workers are sorting the potatoes into large pink sacks. The potatoes themselves should be red, but in this photo they have a white cast. They are frosted with the insecticide Diazanon, which protects them from being eaten by insects during storage. A teenaged worker sits on a partially filled pink sack sorting Diazanon-frosted potatoes by hand. He is wearing a blue baseball cap and a pair of shorts, nothing more.

I shot the photo and backed away, brushing past a pile of boxes of Temec, another insecticide, stacked and waiting for application after the seed potatoes were growing. Then I went outside the shed seeking fresh air to breathe while I prepared for a tense task. I always feel bad for the translator at such moments like this, especially in this case, because my translator was my host in Chile, potato breeder Julio Kalazich, a terribly nice guy. Next to Julio stood Luciano Dallaserra, an efficient-looking fellow in his late thirties and the farmer in whose employ a youth was poisoning himself a few feet away. Dallaserra was smartly dressed, and he sported a nice watch and a blue baseball cap that read CYANAMID. A cell phone protruded from his pocket. He was, by Chilean standards and anyone else's, a prosperous farmer, the largest and most progressive potato farmer in the area around La Serena, which is dominated by prosperous commercial potato farms. He'd farmed this land for eighteen years, following his father and grandfather into these fields.

For a while I let the conversation ramble: farming, business, the virtues of red potatoes over white. Finally, I gave the subject a glancing blow: Did Dallaserra ever worry about the effects of the chemicals?

I didn't need to ask any more directly. He went straight to the core of the issue by pointing out the boy himself: What can one do about such things? He gives them gloves and masks, but they won't wear them, not in this sun, this heat. The boy was an exception. The rest of the workers were wearing gloves, but not much more. Later I talked to farmers of smaller operations who themselves wear no protection against the chemicals, even though the Spanish word they use for them is *veneno*, poison. Every farmer sprays and sprays. A Chilean farmer douses his potato crop as many as nineteen different times during the course of a growing season. His Brazilian counterpart, who faces even worse insect problems, sprays as many as thirty times in a single season. Several farmers could not tell me how many times they sprayed, or how much it cost them. They simply followed the advice of the insecticide salesmen and sprayed whenever they saw bugs. One farmer, acting as if he had just found lice in his child's hair, pointed out the insects on his crop as a sign he should spray immediately. Then the entomologist with us identified them as harmless to his potatoes.

This situation is not unique to Chile, or even to South America. In one sense, it is the result of five hundred years' worth of realizing the power of the potato. Here, at the edge of the central Andes, it was domesticated by the ancestors of the Incas as long as ten thousand years ago. The tuber packs such a dense punch of carbohydrates that it sometimes gets credit for shifting the political shape of

Europe. Armies fueled by the potato have a distinct advantage, as Napoleon's knew. Ireland became so dependent on it that a catastrophic episode of late blight caused what we remember as the Irish potato famine, an event that shifted the demographics of the United States.

Yet during the Green Revolution potatoes were more or less forgotten. The bulk of attention went to cereals, in large part because the problems with cereals are easier to solve. It is a measure of potatoes' sheer productive power, however, that even in the face of relative neglect, they remain the world's fourth-largest crop, ranking just behind wheat, rice, and corn. They are more prone than the big three, however, to attacks by insects, fungi, and viruses. So while growing more potatoes could go a long way toward world food security, as both Asia and Africa are demonstrating, we get a load of pesticides in the bargain. Wouldn't it be great if someone came up with a pesticide-free potato?

Here's another photograph. It has that flash of sci-fi colors and otherworldliness that is the first impression of a lot of microscopic photography, but also a dose of anthropomorphic humor, thanks to the grim countenance of a bug in serious trouble. The bug is an aphid, and it appears for all the world to be wearing a set of six outsized gumboots, which have plastered the aphid to the surface of a leaf as if with glue. The bug is stuck and can't eat the leaf, or anything else for that matter. Or, as this scene is scientifically analyzed by the entomologist responsible for discovering this phenomenon: "The aphids died, and they died because they ended up with these huge blobs of brown crud on their feet."

This is not, however, a scientist's trick; it is a plant's trick, specifically the habit of a close wild relative of potatoes, the discovery of which set the above-quoted scientist, Ward Tingey, and a colleague at Cornell University off on a chase that consumed fully twenty years of their careers as well as twenty years' work from a couple of dozen doctoral students and post-docs now spread around the world busily sticking bugs to plants.

Tingey is a lanky, affable man originally from Wyoming. He's as likely to talk about his annual mule-deer hunts back West as about his science. In 1978 he read a paper a British entomologist had prepared for his doctoral dissertation. While working with some wild tuber-producing plants in the Bolivian Andes, where potatoes originated, the entomologist had come upon one with a curious ability to trap a variety of types of bugs on its leaves. It seemed to be some sort of defense mechanism. The source of this trick was two types of tiny hairs on the leaves, structures called "trichomes."

Tingey was intrigued. He contacted the U.S. Department of Agriculture's germ-plasm bank for tubers in Sturgeon Bay, Wisconsin, which had the species in question on hand, saving him a trip to Bolivia and two or three years' wait while the plant cleared quarantine. (Potato species carry all sorts of commercially disastrous viruses and fungi—remember the Irish potato famine—and U.S. agricultural officials are enormously sensitive about importing them.)

The seed performed exactly as the British entomologist advertised.

The cases of natural pest resistance we have encountered thus far have all involved exquisite and specifically deployed chemistry: a pheromone fends off a particular insect at this exact time; a protease inhibitor counteracts a specific enzyme in the gut of a particular predator. In this case the mechanism is big, bold, and blunt, almost mechanical. You don't have to be a biochemist to understand the process. It works like flypaper. The bug gets stuck. More important, because it is so blunt, Tingey suspected its effects would be general, that it would work across the broad interspecies army that attacks the world's potatoes in the same way that flypaper will catch insects other than flies.

Not that biochemistry isn't involved. Ultimately, biochemistry would become one of the more intriguing aspects of this process, literally on the surface.

Tingey did a bit of poking and prodding on the plants and discovered two types of trichomes—long ones and short ones—that work in a separate and curious fashion. The short one is not so much a hair as a tube full of a viscous fluid. When the mouths and leg parts of insects bang up against this trichome, it ruptures. The viscous fluid spills out, oxidizes, and turns brown and sticky, like molasses—the gumboots on the photogenic aphid. The other type of trichome, the longer one, also secretes a liquid, but it is not nearly so sticky. Instead, it coats the insect's body parts and causes them to rupture the short, sticky trichomes more easily.

It turned out, though, that this sticky business was but the beginning of the wild potato's arsenal. The tuber also secreted a chemical that appeared to have no role in sticking, until it was identified as a virtual copy of the pheromone aphids use to signal alarm.

"It's a volatile chemical that tells other aphids, 'You'd better get the hell out of here,' " says Tingey. And the aphids do.

The combined effect of this list of tricks appears to be an almost total defense system, conferring resistance to a dozen different potato pest insects, a list that covers the potato farmer's dirty dozen: various species of aphids, potato beetles, leaf hoppers, flea beetles, spider mites, and thrips. This is the same list that causes the world's potatoes to soak up more insecticides than probably any other food crop. In 1978 the trick was known to only one wild tuber species isolated in the Andes and to a biodiversity bank in Wisconsin. It would remain a scientific curiosity unless it could be taught to potatoes. There was reason to think the potatoes could learn. Although they lack the chemistry for making the sticky liquids, many domestic potato varieties have trichomes, the structure that is the basis for this defense. Perhaps this was a remnant of a trick they once could perform but which had been bred out of them since they left the Andes.

Sitting at his office conference table as Tingey explains the situation is Robert Plaisted, a soft-spoken man now retired from a plant-breeding career at Cornell. The two of them grin as they recount the events of twenty years ago, as if they were co-conspirators in sticking all those other scientists to what became a sort of mini-industry centered on their two labs. Tingey recruited Plaisted, who realized that if this discovery was ever to amount to anything important, it would have to be bred into domestic potatoes.

Potato breeding is no simple matter. Potatoes are what breeders call tetraploid, meaning they have four sets of chromosomes. Most plants have only two, which makes breeding a crapshoot with a pair of dice. Putting four dice in the game raises the odds against breeding for any one trait to a bet most gamblers would avoid. This in part explains why most commercial potatoes are not grown from "true seed," or botanical seed, as any back-yard gardener knows. Potato plants the world over grow from pieces of potato, by vegetative propagation. That process yields a clone or exact genetic copy of the parent potato plant, ducking the gamble inherent in breeding with four sets of chromosomes.

Worse, the wild tuber in this case is diploid, meaning it has two sets of chromosomes. The necessary cross would not occur unassisted, in ordinary and "natural" circumstances. Fortunately, the wild tuber has the unusual habit of forming male gametes with two sets of chromosomes. This curiosity could, one might speculate, be an aberration on the way to the tetraploid potatoes, a marker on the evolutionary path, but in any event, it was a lucky break that made crossbreeding easier.

Some of Plaisted's early crosses yielded potato plants with trichomes and all of the attendant secretions, but which also picked up a lot of undesirable traits—disagreeable odor and tastes, for example—from the wild relative.

"It takes a lot of selection to get rid of the uncommercial traits that are in the wild species and just transfer the parts we want from the wild species. It has traditionally taken about twenty-five years in every breeding program so far," Plaisted says.

So the trichome-studded potato plant grew into a mini-industry at

Cornell, pulling in two dozen scientists over the course of what became a twenty-year project. By the spring of 1999, organic growers in New York State were producing a commercial variety that resulted from Plaisted's crosses. He demurred a bit about that particular variety; it still lacked some of the qualities of taste and color he and growers would prefer. In his test plots, however, he was growing a better candidate that he endorsed wholeheartedly and planned to release to farmers that same year.

The immediate benefits are easy to calculate: it costs between $60 and $200 per acre per year to spray potatoes with insecticide. Meanwhile, a grower in upstate New York typically gets about $6 for a hundred pounds of conventionally grown potatoes. The organic market pays $30 a hundredweight for pesticide-free potatoes.

Broader benefits become clear as well. Farmers are engaged in an ever-escalating warfare with insects that are constantly evolving to outstrip the evolutionary selection pressure posed by the insecticides. Tingey says that one new nicotine-based insecticide that was highly effective when it was introduced only a couple of years ago is already failing to control evolved resistant populations. That is one of the big points of promise for the trichomes. A doctoral student in the Cornell program bred twelve generations of one pest, each generation from two parents that had survived on trichomed potatoes, a procedure that imposes maximum pressure for adaptation. "He found virtually no adaptation," says Tingey. "It's a mechanical thing in part. It's complex, and it's likely to be fairly durable."

Upper New York State is one matter, but more is at stake here than organic potatoes for health-conscious and privileged urban markets. Some of the experiments at Cornell have been carried out in strict quarantine because their principal actor was an exotic pest, a bug that does not plague temperate region crops, and so is not normally found in New York. (The quarantine is designed to keep it that way.) The potato tuber moth poses a problem for the tropics, where, in recent years, the potato has extended its reach and popularity as a highly productive, low-cost fundament of nutrition. In the Mediterranean region, Southeast Asia, Australia, New Zealand, South America, especially Brazil and Chile, the potato tuber moth is the leading problem, the primary cause of as many as thirty applications of insecticide a year in Brazil. The experiments at Cornell were, in fact, successful. The trichomes do ward off potato tuber moth. So, then, can we count a global problem solved?

Chileans eat potatoes the way Thais eat rice and Mexicans eat tortillas, which is to say a meal is not a meal without them. While standing on the narrow coastal plain that fronts this narrow strip of country with a span from temperate to tropics, one could easily assume that the basis for a cuisine rooted in tubers is always clearly visible: the Andes. Potatoes began here. While some people disagree about exactly where, whether in Bolivia or on an island off Chile, there is no doubt that the Andean region is the source. The conquistadores made room among the piles of looted gold to haul a few potatoes back on their first return trips to Europe, and as a result Europe has never again been the same. Spain robbed so much gold from the New World that it tipped the whole European economy, causing inflation and disruptions that would permanently shift the balance of power. But so did the potato.

The potato made its way somewhat slowly through Europe. The initial germ plasm probably came from Peru, close to the equator, where the varieties evolved to set tubers during short days. In the temperate regions of Europe, however, they mostly failed to produce, because by the time the days became short enough, it was too cold. With its mild maritime climate, Ireland was the exception.

As a consequence, potatoes remained not much more than a curiosity in most of Europe until the late eighteenth century. By then one of two possibilities had occurred: either selection had produced a variety that set tubers during long days or potatoes that already held that trait from the temperate regions of Chile had finally made it to Europe. Still, as we know, Ireland's early experience with the potato left it almost completely dependent on that crop, with disastrous results. In 1845 and 1846 late blight spread throughout Europe, probably imported from North America. In the famine that followed, a million people died in Ireland, and thousands of survivors emigrated to North America.

As a result of the potato blight, plant breeders (the field existed even then) began mounting efforts to identify blight-free varieties. An important factor in avoiding blight turned out to be planting potatoes that ripened early. In the United States, the Reverend Chauncey Goodrich established one of the world's first potato germ-plasm collections in Utica, New York. The U.S. consul in Colombia sent him potatoes he had gathered in a market on the isthmus of Panama. These samples turned out to be varieties from Chile that had evolved the trick of setting tubers on long days. These potatoes became part of the foundation of the U.S. crop.

All this history describes a series of great global circles with a long precedent for connections between upstate New York and Chile. Easy then to stand near the shadow of the Andes now and believe one is returning to some sort of primeval potato, untouched by cultivation and breeding. This is not the case, however. Chile is indeed enamored of its potatoes, but they are every bit as indigenous as are the slightly incongruous names of Santiago's streets, like Bernardo O'Higgins or, for that matter, the name of one of the country's leading potato breeders, Julio Kalazich. The son of a Hungarian immigrant, he grew up on Tierra del Fuego, that last outpost of habitable landscape before Cape Horn. His father believed in potatoes and advised Julio to stick with them, because they would do right by him, and they have. They took him to Cornell, where he became a student of Plaisted's. Once back in Chile, he hooked up with a friend from grad school, a student of Ward Tingey's, Felix Franca, who later returned to Brazil. The two mentors and two students formed the basis for a partnership that put together a McKnight project.

Plaisted has indeed bred a variety of pesticide-free potatoes that will work commercially in upstate New York, but teaching this trick is not simply a matter of spreading that New York variety around the world. There is the matter of local conditions. Chile is ideally suited to lay the groundwork for spreading the innovation worldwide. First, the country lies on a narrow north-south strip of land that provides short tropical summer days in the north and long temperate ones in the south. Brazil has similar climate conditions but also adds some new pests to the mix. Between those two countries and New York, a sufficient span of conditions exists to breed and test varieties that will work anywhere in the world.

Beyond this, Chile has a firm and established relationship with potatoes, not only indigenous ones but also those drawn from the experience of its large population of European immigrants. European varieties are grown here in the center of origin of potatoes, but European varieties with a Chilean twist also rooted in blight and perhaps Kalazich's biggest challenge.

Chileans eat pink-skinned potatoes. They call them red (the name for the leading variety is "cardinal"), but they are really a sort of a pale yellowish-pink. At one time, this habit was simply a matter of convenience, but then an event cemented it in place. In the 1940s a round of late blight swept the country, destroying the crop. By then, blight-resistant varieties were readily available, and a German white potato came quickly to the rescue. Chileans had no real problem with the color, but the white potato turned to mush when boiled. Chileans prefer boiled potatoes. By the time someone got around to developing blight-resistant pinks, the country's consumers had developed a strong and deep-seated resistance to white potatoes. It means that a potato that will sell in New York won't be popular on shelves in Santiago. This fact is not unique to Chile. Capricious market forces, squishy issues like taste, drive the production of food worldwide. So Kalazich and his colleagues put together a proposal to breed locally acceptable traits like pinkish skin color and firmness into the trichome-bearing foundation and, in the process, create a template for introducing pesticide-free potatoes worldwide. Tailoring food to specific local preferences vexes researchers everywhere; they would rather deal in more concrete issues. The one-size-fits-all monoculture of the Green Revolution was the path of least resistance, but too often it ignored the inescapable truth that actual, edible food is the product of all this technology. What is more appropriate for governing the production of food than taste?


On first sight La Serena seems aptly named. In Chile's arid north, it perches on a low shelf over a bay in the Pacific, sun-bleached seacoast under blue skies next to blue ocean. Together with a sister city, it wraps around the bay and is home to more than 200,000 people. The Elqui River runs quick and straight from the Andes, laying out a broad alluvial plain as it slows near the sea at La Serena. The lower, flat stretches are set in farms; in fact, this is one of Chile's most productive agricultural valleys and the primary producer of its potatoes. Upstream, the Elqui's floodplain pinches narrower and steeper, and the fields give way to terraced hillsides of grapes. This is the center of Chile's grape-growing region, producing fresh table grapes in austral summer for the winterbound markets of the Northern Hemisphere, especially the United States, but also for *pisco*, a fortified wine and a sort of national drink. Wine, seafood, and sea provides a setting in which one could almost predict that the town of Vicuña, near the headwaters of the Elqui, set in sharp Andean peaks, would have a special sort of produce of its own. In fact, it is the hometown of Gabriela Mistral, the poet who won the Nobel Prize in 1945. The one-room school where she taught can still be visited today.

We also found Luciano Dallaserra's bare-skinned workers sorting Diazanon-laced seed potatoes in this valley. *Pisco* and poetry notwithstanding, the prosperity that supports the serenity and provides much of the nation's considerable potato crop is heavily based on monocrop industrial agriculture and the pesticides that entails.

Federico Merin, an agronomist who has worked his whole career in the valley, is hard put to say which type of insecticide dominates here, only because the local practice involves spraying a mix of three or four at a time, an application the farmers call *la bomba*. Typically, they spray a dozen times a season and, at peak outbreaks, every four days.

"*Hay problema*," Dallaserra himself tells me.

Two plastic crates of potatoes in the back of a pickup truck provided a centerpiece to my conversations with these farmers. Researchers in the project had dug them that afternoon and were in the process of carting them around the Elqui River Valley for some expert opinions. It was the first time these particular potatoes had seen the light of day. They had come from a test plot operated by the Instituto de Investigaciones Agropecuarias, or INIA, Chile's national agriculture research organization. The potatoes were the progeny of Tingey and Plaisted's work at Cornell, passed through Julio Kalazich's breeding program at Osorno, 1,000 kilometers to the south in Chile's temperate, wet zone. Breeding is carried out there because the area is relatively pest- and disease-free, but once a variety is far enough along for its trial in the real world, it comes north to La Serena, where an environment of monoculture industrial agriculture provides the acid test: a whole array of insects, fungi, viruses, and critical farmers and consumers. One of Kalazich's varieties had come far enough along for a field test. It had grown through the summer and was ready for digging in early March.

A fat, heavy hoe turned the loose, volcanic soil of a field that had gone a whole season without pesticide, and to light came tubers, pale white with a faint red blush, skins unblemished by the tunnels of tuber moth larvae. A row over, the hoe dug conventional cardinals, their redder skins perforated everywhere with heavy moth damage.

"I'm very, very happy with what I am seeing here," Kalazich says. "This is our goal."

Success? Not yet, he says. He's still skeptical about some traits—not red enough, the taste still too bitter—but this variety is a long step in the right direction.

We have the two crates of potatoes with us when we pull into the farm of Hernan Rivera. Like Dallaserra, Rivera is a large grower, one of a handful of major farmers who dominate the region's production. (As a result of land reform a generation ago, this region of Chile consists mostly of small commercial farms of maybe thirty acres. Perhaps one in a hundred will be an exceptional large farm of several hundred acres.) I want to show Rivera the potatoes right off, but he has other matters in mind. He wishes to talk with Patricia Larrain, the entomologist who oversees the potato work at La Serena. Personable, confident, and smart, she seems to fit easily into the work with farmers. Over the years she had worked with Rivera until she won sufficient trust to try an experiment in one of his potato plots. She studded the field with sticky insect traps and taught Rivera how to collect insects from the traps. The sticky traps are not for control but for monitoring the type and number of insects present in the field. Rivera sprays pesticides only when the number of insects in the traps exceeds a certain level which Larrain has determined corresponds to real damage and yield loss. The field also holds pheromone traps, the hormonal lure that inveigles breeding males out of the population in order to reduce some losses. The net result is that Rivera applies pesticide at about half the rate of a typical farm, which has saved him $350 per hectare this season.

Dallaserra has no traps on his farm, but when we visit him, Larrain slips off to look for the telltale signs of a progressive farmer, and she finds them. He has the latest in irrigation. He has planted sunflowers to attract pollinating bees to his melons. He is experimenting with a variety of crops to break the potato cycle. A farmer who experiments is usually open to more experiments, so Larrain talks to him about trying bug traps. He thinks it's a good idea. He'd like to try. This is how Larrain responded, in a measured way, to seeing that bare-skinned worker a few meters away. Rivera and Dallaserra are successful commercial growers, but also early adopters, farmers who are forever experimenting with new ideas and technology to stay ahead of the curve. They provide the edge where people like Patricia Larrain choose to apply their levers in the hope that less-prosperous neighbors will watch these farmers and follow their lead, one avenue of change in farming culture.

Interestingly, before our visit to Dallaserra, Julio Kalazich was confident about his potatoes' ability to survive bugs but not market forces. He didn't think them red enough. That's probably why he seemed a bit incredulous after hearing both farmers' opinions of their marketability. "They think they might work," he says.

Both farmers would be willing to try field tests. They say it is simply not true that Chilean consumers are more resistant to eating red or yellow potatoes, especially if those consumers can be assured that the odd colors mean the potato was produced without pesticides. Most of the produce from these farms heads south to Santiago, a cosmopolitan, upscale city full of malls and supermarkets such as one might find in Palo Alto or Bethesda. Consumers in Santiago are open to something new. In fact, they are already buying yellow potatoes from Argentina. The resistance comes from hidebound wholesalers who persist in believing only pinks will sell. This is the sort of precise information that helps researchers form a strategy for overcoming market resistance. Researchers must work with wholesalers, not consumers, which requires a different set of tactics.

The farmers' assessment has Kalazich scratching his head, not because he doesn't believe the information, but because the success of his breeding operation now comes down to the capriciousness of a few wholesalers and the necessity of unleashing market forces to overcome that resistance. What does a breeder know about this business? Not much, but he allows he'll have to learn at least enough to bring the right people to the project. All this research has to become interdisciplinary. Lots of good crop varieties have been bred and now languish as "on-the-shelf" technology. By now breeders like Kalazich know their job is not over until the market has accepted their work.

The farmers also told us that they now grow potatoes only in the cool winter months, and in summer grow grain and other crops under irrigation: maize, sorghum, fava beans, melons, celery, peppers, wheat. This sort of diversity is very new and not at all common on other farms in the area. The more usual practice is to grow potatoes after potatoes, because potatoes are still the most profitable crop. Both of these farmers tried and liked rotation, not so much because of the economic diversity but because their potatoes became even more profitable when they practiced rotation, which gave them better yields by improving the condition of the soil and decreasing pest problems that resulted from monocropping. They read. They talked with experts. The experts suggested rotation to them. They tried it, and it worked.

The danger of magic-bullet solutions like the trichomed potatoes is that they may once again encourage monoculture by defeating the pests that encourage rotation. But other pests—nematodes, blight, and viruses—won't be affected by the trichomes. Even with the resistant potatoes, rotation will still make sense, and the project has already begun gearing up for an information campaign to ensure that their solution doesn't trigger more problems by discouraging rotation.

The necessity of educating farmers has been a lesson the Chilean government has learned the hard way. Not so long ago changing the minds of farmers was not such a big job, simply because there were far fewer farmers. Incredibly, before the land reforms of the 1960s, about 100 farmers owned all the land in the valley, the same amount of land now shared by thousands. Then came the socialist government of Salvador Allende, and land reforms with it. No one I talked to in the Chilean project had opposed Allende, and some still labeled themselves socialist. One of Kalazich's closest colleagues was forced to leave the country when Augusto Pinochet, with the help of the U.S. government, staged the coup that killed Allende in 1973. Kalazich's father-in-law was imprisoned. Still, these same people are frank in their assessment of the results of Allende's effort to redistribute land to poor people.

"I think it was a failure, because small farmers didn't know how to handle production," says Carlos Quiroz, director of the INIA region headquartered in La Serena. "They didn't have the skills. It was a complete failure."

Only a few years ago, a single field day was held for demonstrating new techniques to farmers in Quiroz's entire region; last year there were fifty. One small farmer proudly showed us the new tractor parked prominently in front of his modest little house. He had bought it with a government loan administered through his cooperative. That same loan buys quality seed and chemicals, but also comes with strings attached: The farmer must hire a consultant, someone paid to advise him on how to get the best yields. The second Green Revolution, the one now brewing in similar projects around the world, can be called an information revolution on two fronts: the raw information contained in genetic resources and expressed through breeding and genetic modification, and also cultural information flowing between farmers and researchers. Increasingly, we are coming to realize that one is useless without the other.

The front office in INIA's La Serena building is a particular prize, not so much for Quiroz's authority as for the view. It sits on a hill overlooking the city and the sea beyond. It is one of a clutch of neat buildings wrapped in a courtyard that Patricia Larrain and some volunteers have landscaped with shrubs, flowers, and trees. Across the courtyard stand a few sheds where she and her assistants do some of the experiments with bugs. One could get easily lost in the scenery and miss the significance of what goes on in her plastic boxes.

If the trichomed potatoes do deserve some sort of magic-bullet status, it is based in the mechanical nature of the process.

"It's a trap, a sticky trap. The aphid dies of starvation," says Larrain. "In the end it's glued to the surface of the leaf."

Entomologists such as Larrain like this scenario so much because it's hard to imagine insects evolving resistance to overcome it. Maybe in the end the biggest problem with pesticides is that they force evolution of resistance, a constant problem now in La Serena and worldwide. It does not, however, matter a bit if the chemicals come from the end of a nozzle or from the plants' own sophisticated biochemistry or is induced by genetic engineering. Plants genetically engineered to produce the "natural" and harmless insecticide made by the bacteria *Bacillus thuringiensis* are already producing strains of insects immune to Bt. Decreasing the world's reliance on pesticides does not mean we are decreasing our reliance on chemistry, so it's still an arms race. Although not so with the mechanical trichomes, or so it seemed.

Larrain's plastic boxes contain what entomologists call "no-choice tests." It's one matter to have a resistant plant in a field of other plants; the pest can simply move to another plant. An insect enclosed with only a resistant plant has the choice of eating the plant or starving, thus "no choice." Potato tuber moths avoid trichome material in the no-choice tests, but the only material in the bug box is the tuber. In fact, there are no trichomes on tubers, only on its foliage. Still, most adults avoid landing on the tubers. Furthermore, only a few of those who do will lay eggs, and only a few of those will survive. The resistance resonates through adult, egg, and larva.

Researchers ask this question of the tubers because they know the habits of the tuber moth; that the adults normally lay eggs on the foliage, making trichomes effective there, but some will lay eggs in the soil. The moths could circumvent the trichomes and beat the game if the tubers themselves had no resistance. So is the resistance in the tubers good news?

If there are no trichomes, no mechanical structure to explain it, then there must be a chemical component that researchers don't yet know about. Genetic mapping has already told them that the whole phenomenon is complex, that it is printed out through a series of genes. So there may be a separation of the chemical and mechanical sides. Maybe they can work independently. Maybe not. Whatever the chemical basis, it's in the tubers themselves which humans eat. It's not at all unusual for plants to be laced with chemistry that insects find offensive and humans don't. Potatoes are notorious for being laden with all sorts of foul alkaloids that are either rendered harmless by cooking (ever eat a raw potato?) or that human body chemistry can counteract. The dining room at Kalazich's research station served us trichomed potatoes for lunch and they tasted like potatoes, but who knows, maybe there was a good reason that the wild ancestor that made trichomes was left wild by the domesticators of potatoes. A researcher's answer always tends to lead to another question.

A development as sweeping as Ward Tingey's work with trichomes tempts us to give a great global sigh of relief that a major problem has been solved. Maybe that is so; it's hard to imagine a single simple step more important to the world's agriculture, the environment that contains it, and the nutrition that depends on it. Powerful tools provoke what is often called the hammer principle, which says that when you first discover the hammer, everything starts to look like a nail.

We are in Osorno at the INIA research station where Julio Kalazich is based. Osorno is a prosperous farm town in the south's wetter climate. All around roll gentle hills, a smattering of trees, dairy cattle, and small, tidy farms such as one would expect to find in the upper Midwest in the United States. We walk a path with Kalazich through row upon row of potatoes of every variety, trichome-bearing and otherwise, wild varieties and commercial, even some spindly, barely domesticated potatoes that are very close to what the Spanish first took with them to Europe. Some of the plants have already been dug, exposing neat sorted piles of potatoes of various colors, to be weighed, measured, counted, cooked, and tasted by human and bug. Kalazich is not yet happy with the varieties that have made it to La Serena and the real world. More are in the pipeline that he likes better, but he's still not there, and as he walks the path through the field handling tubers he looks first for a reason to reject them, he says. All these rows, though, seem set on that single path of trichome development. But, in fact, there is no single path.

Our group has just been in a discussion, a heated one really, something of a confrontation about insecticides and potatoes. The catalyst was a particular woman who is nothing if not catalytic—Rebecca Nelson, a potato scientist with the International Center for Potatoes in Lima, Peru. Nelson is outgoing and direct, with a disarming, warm face, a young mother approaching middle age, a MacArthur Fellow, a never-met-a-stranger American who is as blunt as that breed can be. She was there to evaluate the program on contract with McKnight.

"We are at a point where we are beginning to change," said Patricia Larrain, whose nonconfrontational approach to this matter we have already seen at work. Nelson asked the researchers about human health problems from the pesticides.

"There is not much concern in the public," said Carlos Quiroz. "The problems are difficult to prove."

At which Nelson more or less explodes. "The chemicals are really nasty and people are swimming in them."

This outburst prompts an exchange of anecdotes. All the people in this room have seen such scenes as pregnant women working among tall, freshly sprayed potato vines, dripping so many chemicals that the women looked as if they had taken showers. Or workers stirring pesticide solutions with their bare hands. Still, researchers want to believe that their breeding program will solve the problem quietly, avoiding confrontation with chemical companies by eradicating the need for pesticides—in spite of the copious evidence that the average farmer's information, not just in Chile but in all the world, comes from pesticide salesmen.

"The chemical companies are going to harass you, but that's your job," says Nelson. "Your technology is going to cost them money. I have said some scandalous things about this, and it's going to get me in trouble, but somebody has to be militant."

She insists that simple plant breeding will not be enough in this fight, that these researchers who were trained only to breed potatoes will also have to fight a propaganda war, an economic war, a legislative battle in a country still shrugging off the effects of over two decades of military rule. It's not so much that the Chileans are offended or that they disagree with what Nelson has said, but more that they are daunted. Their training and expectations lead to a search for scientific solutions, but the solutions now seem to lead to another set of problems much more difficult to grasp.

Later Kalazich sets up a side trip to another field, to Pepe's field. It has nothing to do with trichomed potatoes but complements the notion that we need a broad arsenal of solutions. Pepe is José Santos Rojas, a potato breeder who specializes in seed production. He was Kalazich's first friend at the station in Osorno when Kalazich joined the staff in 1979, in the days when Pinochet was still in power and certain subjects were not discussed.

Santos's field is much smaller than Kalazich's expansive menagerie of potatoes, and serves an altogether different purpose that can best be understood by looking again at the bare-skinned worker sorting Diazanon-dusted seed potatoes. The key part of the image now is seed. Because of vegetative propagation, about 20 percent of each year's crop is saved to seed the next one. Just like tubers in the ground, these potatoes supply carbohydrates alluring to an array of attackers—insect and otherwise—and must be protected during this microcosm of the growing season with an array of pesticides. That is to say, the seed of potatoes, not the crop itself, accounts for a significant share of the pesticide load.

Worse, the protection doesn't work well against all insects, or against other attackers such as viruses. Seed potatoes are living tissue, so they carry disease from season to season. In most regions of the world, disease-free areas are set aside for growing seed potatoes, an expensive but unavoidable precaution. None of this is necessary for crops that propagate by seed, a fact that partially explains why cereal production feeds the world for the most part. It is only because potatoes are such a productive crop that they still manage to hold down the number-four spot in the hierarchy of key crops, despite the hurdles.

Pepe's field could improve that standing. It is producing true or botanical seeds, seeds like those of any other plant. The trick is not new; germ-plasm banks have long depended on true seed to move disease-free potatoes around the world. The process, however, has heretofore been so difficult as to be reserved for extraordinary circumstances. Thanks to the potato's four sets of chromosomes, open pollination is a genealogical gamble that can easily lose traits carefully bred into lines. Santos's group of breeders has found, however, that by carefully assuring the lineage of breeding stock within families, they get close enough control to assure reasonably uniform progeny. They hand-pollinate the potatoes just before the flowers are mature, before bees are attracted which would confuse the issue by bringing in pollen from other sources. The operation is not high-tech, just meticulous, and therefore appropriate technology.

As we toured the field, we were joined by an enthusiastic and articulate ag researcher from the Universidad Nacional Autónoma de Mexico in Mexico City. Jorge Sarquis had participated in a true seed project in Chihuahua State in Mexico in which a destitute village was taught, and then successfully used, the technology.

"We are past the point of demonstration," says Sarquis. "We are helping to feed 50,000 people who before were dying of cold and starvation." More than pesticide use is at issue. Using methods Santos has developed, it costs only about $1,000 to produce a kilo of seeds, enough to seed 10 hectares of crop. Seed potatoes for the same area cost $6,000. Meanwhile, the seed potatoes normally saved for seed can instead be eaten, instantly boosting the average yield by 20 percent.

The technology consists of simple screenhouses, a breeding line, and some relatively basic information.

"This is food security," says Kalazich.

# In Wildness
# Is the Preservation of the World

## Sustaining Traditional Farming and Genetic Resources

### [Mexico]

Even if a visitor knew no Spanish, he could have listened to the lilt-ing conversations in the countryside around Mexico City during the summer of 1998 and learned the words *triste* and *niño*. Ultimately, the situation then under discussion in every village may indeed work out to be about sad children, but the farmers were actually describ-ing the pathetic state of their crops: *triste maíz*, sad beans, sad squash. The child in question was El Niño, and every farmer knew this shift in Pacific Ocean currents was chief among the forces that had sent a despotic drought that endured even past Corpus Christi Day, into June, the usual heart of the wet season.

Of course, as in any catastrophe, some farmers will do better than others, the issue throughout this story. Surviving adversity is the core of natural selection for both farmers and crops. Mexico is a better place than most to begin considering selection, a fact that becomes altogether clear in walking the tidy maize fields of Manuel Montes de Oca. Montes is one of the farmers who will do better than the others, which is why a group of scientists from Chapingo University, Mexico's primary agricultural school, and from other universities are in these fields.

To describe what is at stake in this field will take the whole chapter, but a good place to begin is by thinking about a plant breeder. Most of the gains in plant breeding have come from trying to cultivate an array of varieties, or races, of a given plant species under various conditions, then selecting those that perform well and cross-breeding them with varieties that have other desirable traits in order to pass the information encoded in their genes to succeeding (the double meaning of the word here applies) generations. At one time it would have been enough for university professors to hand the resulting seeds to farmers and say, "Go and prosper." But we are coming to realize that farming is a system in which the seed is literally only the kernel. In the briefest of shorthand, let's imagine that researchers begin treating whole farms like individual plants: they see which farmers perform well, then spread these farmers' information to other farmers, an upscaling of natural selection. Think of a test plot of farmers, not of varieties. Researchers are screening for desirable characteristics in a whole farm system, not just in a given crop variety.

Farming is not so much a seed or even a brand of technology as it is a process, a system, and in Mexico there is a name for one such system, the "milpa," an ancient method of growing maize, beans, squash, and an array of leafy plants called quelites in combination. The scientists in Montes's fields are part of a program headed by Universidad Nacional Autónoma de Mexico, or UNAM, specifically to study *in situ* conservation of the milpa system, a program fraught with meaning for the rest of us, down to a rethinking of what we mean by conservation.

Montes is a young man, just in his twenties, thin and quiet. He is demure when introduced to visiting researchers and a journalist in his village, San Cristobal Poxtla, a neat rack of adobe and concrete houses set in the Chalco Valley an hour south of Mexico City. A few Holstein dairy cattle wander the streets, dogs cavort, people go about their business on foot. Everybody *buenos días*es everybody else.

The scientists introduce Montes as an agronomist, meaning he has had formal training at one of the nation's agricultural schools, two years beyond normal high school work. Serious preparation for assuming a role in the family farm. Along with his father and four brothers, that is to say, six families, he pulls a decent living out of four hectares (about ten acres) of ground they own and another five hectares they rent in most years. They have a small used tractor and twenty milk cows. They grow maize and beans as cash crops and raise an array of fruits and vegetables for the family to eat. All this places Montes in the middle in terms of both prosperity and methods in a national agriculture that includes everything from large-scale corporate monocrop operations to subsistence plots farmed as they have been for thousands of years.

Reticent to admit it, Montes realizes that his maize is doing better than his neighbors'. Yes, the drought makes him sad, but not so sad. I press him as to why and he doesn't want to say, maybe a mix of country humility and scientific skepticism. He will say only that it appears to be doing better, but it's too soon to tell for sure. The experiment is not yet over. Everyone present thinks it's because of the manure from the dairy cows. Montes has not yet switched to total monocropping and the complete reliance on chemical fertilizers that comes with it. The manure from his Holsteins retains moisture in the spaces its rotting makes in the soils. Chemical fertilizers do not, and this extra moisture may be the edge his corn needs, a bit of the selection.

The manure leads us straight into a discussion that destroys any illusion that this village is isolated from world events. The acronym NAFTA can be heard in a Mexican farmer's conversation almost as often as drought. The rules of the game have shifted, and the more open markets that resulted from the North American Free Trade Agreement with the United States and Canada have brought in a glut of cheap American corn. Mexicans don't like it. Connoisseurs of maize, Mexicans grow a stunning array of colors that sort roughly into categories of white, yellow, black, and red. White for tortillas.

Fat, monster yellow kernels for the corn stew *pozole*—a national obsession. Mexicans regard American corn from the Midwest as big, yellow, and tasteless. Editorials rage against the tortillas it makes, as they lament the fact that Mexico is forced to accept it because of a combination of international trade policies and failed agricultural policies in their own country.

But to Montes, the maize farmer, NAFTA means something else. It has also allowed the importation of cheap milk, and the income from his Holsteins has fallen sharply. No sale of milk, no cows. No cows, no manure.

The industrial model offered by these global markets favors simplification, but there is another side to the story. Montes takes us now to a second field, also of maize, but of blue corn this time, and it, too, is doing tolerably well, given the drought. Mexican farms have always grown blue maize. Montes likes to grow it because he likes to eat it. In recent years, though, it has become a significant cash crop, thanks to a strong demand for blue corn chips, both domestically and internationally.

He reports that buyers now come from all over the world to his village asking if anyone grows blue maize. Given the premium price involved, it has not taken long for a lot of local farmers to begin growing blue maize.


The marketplace as an institution has been around as long as agriculture. In fact, it has been as integral to the development of agriculture as the selection of seeds and the refinement of farming techniques. The rise of agriculture ten thousand years ago can be seen as a revolution not so much in raising food as in storing food. With storage came transportation, and with transportation came markets, city, hierarchy, and the rest. Before agriculture, people ate what they gathered from the plants living on that gray, blurred line between wild and domestic. Human culture sharpened that line over thousands of years, until finally agriculture stood clearly separate, a technology that included storage and markets.

That the fortunes of a single family in the Chalco Valley should fluctuate with the caprices of markets in Mexico City is nothing new for the people who have grown maize in this valley for more than five thousand years. The markets probably were no more merciful when the Aztecs controlled them. It is also probably true that globalization makes the swings wilder, the troughs deeper and longer, just as global weather changes like El Niño make droughts more severe. But this place rolls with the punches. It helps to consider this while walking Montes's drought-stricken rows. When you lift your eyes to the south horizon, where, weather and the city's smog permitting, you can see the shoulders of two snow-capped volcanoes, Popocatepetl ("smoking mountain") and Iztaccihuatl ("white woman"). They are the nation's second- and third-largest volcanoes. The valley was and is formed by volcanoes, humps, lava flows, and earthquakes. Someone once called Mexicans "sons of the shaking earth." Upheaval is the rule here, and it's difficult to imagine how villages like Montes's have maintained six thousand years of settled life in the face of it. Conservation, *in situ* conservation, at first glance appears to be about insulating a place from the forces that would change it, but here is a place long used to forces. It's a shaking *situ*. Something else can be learned here.

Looking down from the mountains, you might notice that Montes's fields hold a few weeds. Among them are teosinte, a wild grass and probably the ancestor of maize. Growing with it are wild ancestors of tomatoes and beans—more than a reminder of the deep history of the place. The plant breeders of the developed world, when trying to make improved crops, time and again go back to the source, specifically to teosinte, to breed with maize to extract desirable traits. In doing so, these breeders are taking a second run at domestication, as they repeat the very processes that went on in this very valley five thousand years ago to give us maize. It's about reinventing it. Except the teosinte standing in Montes's field suggests that the process never stopped. And given that, where do we draw the line between domestic and wild?

I am walking Montes's fields with Fernando Castillo-González, a professor of plant genetics with a Ph.D. from North Carolina State, and Rafael Angel Ortega-Paczka, a plant breeder with a Ph.D. from Russia's renowned Vavilov Institute. Both are researchers in the McKnight program and teach at Chapingo and the associated Colegio de Postgraduados, a postgraduate school. Those two schools plus UNAM and the University of California at Davis are allied in a broadly based and diffuse project that examines the milpa system. The shorthand description of the project is the one used above—that the researchers are examining *in situ* conservation of milpa; that is, finding ways to make certain it goes on. Milpa is a multicropping system: a series of crops are planted together, yielding benefits lost if the same crops are grown separately. Unlike all of industrial agriculture, it is a polyculture. As such, it offers some clues for breaking out of the bind in which conventional agriculture has found itself. Reading those clues, though, entails understanding each of the crops in the system—maize, beans, squash, and the leafy, weedlike quelites— and the relationships among them, as well as relationships with the larger ecosystem. The research here presents a classic study in system dynamics, in that reductionism will simply obscure the most interesting question, which is: Why is the whole greater than the sum of the parts? But because this system is intimately woven with the culture that has used it since the Aztecs, the research is also necessarily wrapped up in social systems. The questions are so complex that it may be best to stick with the shorthand description of the work for now and let the horizon lines form as the story broadens.

It helps to know, though, that the researchers have at least broken down the range of questions by focusing their work in two separate geographical areas. An explanation of the value of this approach emerges as I walk the fields near Chalco with Ortega and Castillo. I mention to them that although the program focuses on *in situ* conservation of the milpa system, intercropping, we are seeing not milpa but something far closer to the monocrop agriculture of the developed world. True enough, they respond and tell me that is the general rule among the farmers the project works with in the Chalco Valley, one of the two geographical focal points. Later, we will visit an area a few hours away in the Sierra Madre Oriental, where we observe milpa in a purer form, the other end of the continuum. These farms in the Chalco Valley in many ways represent the cutting edge of loss; the others in the mountains, a purer form. A generation ago, the Chalco farms were milpa, but the demands of the markets or of the times or even ignorance are stamping it out. Yet they tell me that the program is not here as missionary to try to stem that tide but more to observe the process, to observe change. To see what survives.

Out front, the milpa project has two goals: to improve the farmers' practices and to conserve the milpa system. Chalco can suggest that those goals are at odds, an enigma that makes Castillo and Oretga grin.

"We try to continue what is. We are here to keep the system as it is," says Castillo.

"We try to improve what is, not change what is," adds Ortega.

Change lies at the heart of a dynamic system's ability to survive upheaval: that which we seek to insulate from change has endured so far only by changing.


Critics of the milpa project and of similar projects around the world suggest that it is romantic to try to save a traditional and outmoded system of subsistence agriculture better replaced by technology-driven methods. The critics suggest that this work is more about museum pieces than farming.

The critics would do well to consider some of the work at Chapingo.

Clemente Villanueva Verduzco, Ph.D, a squash breeder at Chapingo, and I drive one hot afternoon from his campus office to a greenhouse at the edge of the campus. At one edge grows his experiment, a row of starkly green squash set against another row of starkly dead ones, the sort of black-white result that strongly suggests the experimenter just may be on to something. The common treatment of both rows was the herbicide Tordon. Villanueva is breeding herbicide-resistant squash.

The world of agricultural plants is sharply divided into two parts—grasses and forbs. The grains are grasses, and most of the rest are forbs, or broad-leafed plants. Nature and the milpa system grow forbs and grasses together, taking advantage of certain complementary features. Some forbs, for instance the legumes like beans, pull nitrogen from the air and "fix" it in the ground. The grasses then use it as fertilizer. Yet most of what we call weeds are also forbs, so industrial agriculture has taken advantage of the sharp differences in the two classes and developed herbicides that kill forbs but not grasses. Tordon, an artificial growth hormone chief among these, causes forbs to grow themselves to death, leaving corn- and wheat fields weed-free. Herbicides have shown up heavily now in Mexican agriculture, meaning the end of intercropping beans and squash with corn—the advent of monoculture.

One way out of this box is to convince farmers to stop using herbicides, but Villanueva believes a farmer is going to do what a farmer is going to do, so he breeds squash to roll with the punch. He believes this course will be controversial with some people, even American journalists, but he shows me anyhow.

There is a background to his pragmatism. That same day, on a visit to a farmer's field with the rest of a group of scientists, I had noticed that Villanueva's soft-spoken conversations with the farmers seemed to run in a deeper groove than those of his colleagues, a rapport that showed even through the language barrier. So in his green-house I asked him if he had grown up on a farm.

"In my family, we are fourteen," he told me, and yes, all those children were fed on his father's small farm. His town had no high school, but Villanueva got lucky, found out about and successfully wrote an exam that got him into a boarding school, then another that got him a scholarship to Chapingo. Two other brothers got an education, as did a sister. The rest get by, including five who migrated to the United States to find what work they could.

The principal researcher of the milpa project is Robert Bye, a Harvard-trained American ethnobotanist who has taught at UNAM since 1981. He says the trick of running the whole thing has been managing its diversity, that the survival of the milpa is embedded in custom, human choices in the face of global forces, and so of necessity the program has become broadly interdisciplinary. It was enough of a stretch to get squash scientists working with corn breeders, but it soon became apparent that the list would have to include economists, anthropologists, and ethnobotanists, and so it now does.

"Perhaps the whole secret of this is how people respond to change," he says.

The academic diversity mirrors the diversity of the plants of the milpa. The principal characters—maize, beans, squash, and the leafy quelites—are sorted according to structure. The maize has a strong vertical shaft, and most of the beans are climbers, like the pole beans of a back-yard garden. The beans evolved using maize plants as poles, paying back what they gained in support as nitrogen for the corn. The strong vertical element left a horizontal niche which squash runners filled, the two dimensions forming a sort of grid. Quelites, a range of species of leafy forbs, filled in the spaces in between. They provided some ground cover, held moisture, prevented erosion, and probably even served as traps for insects.

The quelites, however, also embed themselves within a social matrix, an even more intriguing issue for the project. Among the Indians who developed the milpa, the quelites were also food that seeded itself and could be gathered throughout the growing season. As monocrop agriculture eradicated these "weeds," people began relying instead on substitutes, such as spinach and lettuce from the markets. The milpa project has done some nutritional analysis of the quelites and found that they provide three to four times the vitamins and minerals of spinach. Yet the sharp prejudice against anything Indian has pushed quelites out of the Mexican diet, a process that began with colonization. One of the common quelites is amaranth, which is now available in trendy American health-food stores. It can be a brilliant red plant, a fact that caused the Spaniards to ban its consumption. Eating red food was held sacrilegious: it evoked the blood of Christ.

"We are trying to slow or reverse the social stigma against eating weeds," says Bye.

I had been talking about quelites with scientists for several days but had not really seen any growing until researchers took me to a test plot. Flanked by a contingent of students, most of them at home in the fields considering that 90 percent of Chapingo's students come from farm families, we walked rows of squash in test plots that could have been at Cornell or Michigan State or any American land-grant school, except they were full of weeds, of quelites. Villanueva showed me one which looked familiar. As a child I'd hoed this very species out of cornfields in the upper Midwest and learned to call it pigweed. Others call it lamb's-quarter. I'd hoed it until my grandfather, the one with the Chippewa Indian ancestry, convinced me to eat some of this weed. He told me it was better than spinach.

Jorge Acosta is a plant breeder in Mexico's national agriculture program, quick-spoken and smart. He talks about his job in easy, colloquial English he picked up while working on his Ph.D. at Michigan

State University. He specializes in and makes a hobby of beans, in that in addition to his regular duties he collects beans from throughout the country. He crosses them with elite lines in an attempt to produce a more insect-resistant variety. In this breeding, Acosta has a choice: he can select from among the thousands of local varieties farmers grow or he can do as many breeders do and select from the wild relatives of beans. Because bean domestication took place in Central America, wild relatives abound; one genus alone offers at least forty species of wild beans in Mexico. Given these options, Acosta chooses neither. He prefers what he calls the weedies, naturally occurring crosses of wild and domestic beans already growing in farmers' fields. Because they have already crossed, generations of his work have already been accomplished.

Maize also grows with its wild relative teosinte, so something similar occurs. The milpa project has surveyed farmers' attitudes about teosinte and found that most regard it as a weed and eradicate it. Some, however, don't mind it so much, and a few, just a few, believe it's important to leave teosinte growing in the maize because it helps breed a better stock.

The concept at work here is gene flow, the continued interbreeding of domestic and wild. Researchers have repeatedly demonstrated this natural process. It exists. It continues.


One of the five farmers with whom the milpa project works extensively, José Concepción Morales Martínez, lives in the Chalco Valley. His house rests comfortably in a classic walled courtyard in the village of San Juan Coxtocan. Just down the street—a visible reminder of the stakes of this work—a government truck pulls up to become a roadside welfare office, distributing food for the village's poorest people. In fields all around, the drought-stunted stands of corn suggest that the truck will draw bigger crowds as the year wears on.

A big house minds the wall toward the front of Morales's courtyard, but it's not the farmer's. It belongs to a relative who has had some success as an opera singer and prefers village life. Morales's house is a modest adobe, but a close-to-new Ford 6600 tractor stands nearby, a better measure of Morales's relative prosperity than the house. A friend extended the farmer some credit to buy the tractor, and a few good years in maize paid it off. Morales owns about fifteen acres and, with rented ground, farms a total of about fifty acres. Like his neighbors, he farms mostly maize.

Morales is not *en casa*, but sits at a patio toward the rear of the courtyard with a couple of generations' worth of family amid a great mound of unshucked ears of maize. Last year's crop is stored in its husks in bins and husked gradually by hand through the year and sold. A small portable TV plays, and a couple of dogs sleep in great piles of dried husks.

Traditional diet makes these husks a sort of informal currency that doesn't show up in formal economic assessments. Even the wealthy of Mexico City preserve the habit of wrapping tamales in corn husks, so small bundles of carefully preserved husks wend their way through the country's markets. Farmers like Morales often pay their workers in husks, so they act as scrip in markets. The economists working with the milpa project have found that the husks produce as much income for these farmers as the maize itself, which helps explain why they persist in the painstaking business of husking by hand. It also advises why farmers often reject the modern, high-yielding seed offered to them: inferior quantity and quality of husks.

Most of Morales's maize consists of white maize, *maiz paloma*, maize for tortillas. But we also find *ancho*, a variety from Morelos that will fetch double the price of other varieties. Also available: red corn and blue corn and the huge kernels for *pozole*. There is a continuity in this, a stability based on varieties grown in the Chalco region for thousands of years, to the point that *chalqueno* corn is recognizable and distinct in city marketplaces. The place is of maize, as are the people.

The maize scientist Rafael Ortega from Chapingo tells me that at conquest the Spaniards took one look at the Chalco Valley and pronounced it wheat ground, then ordered the farmers to plant it all in proper European wheat. Farmers in the area still plant wheat from time to time, geared as it is to funnel more cleanly into international commodities markets than, say, corn husks. Wheat is nonetheless the exception, and it's difficult to say whether it is national pride or pride in maize that shows through as Ortega tells me maize will out.

Almost the same sense of pride comes through when Ortega reports the results of a series of field trials throughout the region. Researchers simply pitted modern hybrid corn varieties against the local varieties, and the locals outperformed the others across a range of conditions. Yet the system is flexible. When corn from other regions showed up in Mexico City markets at twice the size or fetched twice the price, local farmers made sure some of that corn came home as seed.

The farmer talks for a while about varieties, and much of this boils down to use—which works best for this or that—but he also makes it clear that there is no single best. It depends. In a drought year, this does better than that. This one weathers storms better than that, this one holds up against pests. This one he just likes better, and this one's husks shape up better by Christmas time, when the demands of traditional feasts pump up husk prices. Ultimately, then, he favors variety, because farmers who don't hedge their bets against the caprices of weather and markets don't get to own tractors.

A post on the patio braces the ceiling above, the floor of the storage attic for maize. Morales's father sits near it, but a collection of large ears of corn hangs on it—a cross section of varieties, but also many crossbred ears that contain the kernels of several varieties. The collection represents the farmer's seed. During the process of hand husking, farmers always watch for exemplary ears and, when they find one, set it aside for the future. They like the checkerboard-patterned crossbred ears, because each encapsulates the stability of variety in a single package. A maize farmer in an even more remote section of Mexico showed me his seed collection, all of it mottled black-, red-, and white-crossed ears. His collection hung on a post in his house next to the shrine to San Miguel, the patron of his family for five hundred years, he said. He can trace his ancestry in people or in his own ancestors' breeding line of maize, both displayed on the center post of his house.

Part of the milpa project is teaching these farmers methods of natural selection, but the farmers have been practicing natural selection continuously since long before their houses had saints. The plants that support these people are not relics from the past so much as they are the result of a continuous process. Still, the process is being refined, recalling the enigmatic goals of the milpa project. Traditional farmers select, based on seed quality, what they like, chosen at harvest. Researchers select based on what a scientist thinks a farmer might like, and there are legions of stories about the latest-touted variety fresh out of a lab at some land-grant school that either fails the test of real conditions or that farmers simply won't grow because they don't like it. The milpa project uses a hybrid approach.

Castillo says the biggest problem with the farmers' methods lies in the fact that they choose the best seed but cultivate a line without regard to the quality of the plant that raises the seed. For example, native maize plants tend to be terribly tall, more than three meters. This means the plants must put extra energy into raising a stalk. Also, the stalk resembles an unsteady tower ready to topple in a windstorm: lodging. All things being equal, short plants would be better.

The solution has been to integrate a plant breeder's normal mass selection process into a farmer's course of work, to suggest to farmers another sort of selection they themselves will practice. During the growing season, Castillo and Ortega go to farmers' fields and work with the families to pick out individual plants with desirable qualities—shorter ones, for example. At harvest, they set the ears from those plants aside, and from that sample the farmer selects for the best kernels, using his normal criteria—one way of improving without changing the foundations.

The pole that supported both saint and seed also supported the attic in the farmer's house, the area over the living quarters that serves as household granary. Centuries of this practice have spurred the evolution of a range of insects to work this man-made niche; much of crop loss to insects in Mexico occurs after harvest. The milpa project works to find out which seeds are most insect-resistant in storage so that researchers can breed that trait into the varieties farmers favor.

All this rests on a sort of bellwether strategy, a belief that, to effect change across the entire countryside, it is not necessary to convince all farmers to change or try a new idea. Getting a few farmers to do so, as long as they succeed conspicuously, means that the rest will happen on its own.

We were at Morales's farm not to see maize but to see squash, a goal best approached by searching out turkeys. The domesticators of pre-Columbian North America gave the world maize and beans, but also squash and turkeys, not to mention chiles and tomatoes. Turkeys are emerging as an important species in the issue of world food security because they are among the most efficient animals at converting the carbohydrates of grain to protein. A particularly impressive specimen was doing just that, pecking away at corn droppings among the husks, sleeping dogs, and the portable TV in the storage shed at Morales's farm.

The turkeys of central Mexico look mostly wild, not like the bleached-out white of factory-scale domestic flocks in the United States but more like mottled gray-brown megabirds one would expect to see in the forest. The officious toms execute a cakewalk strut, ruffling in full bluster against any human treachery aimed at protein conversion of one of the hens or her brood.

What all this has to do with squash is that turkeys eat squash, as do many of the other members of José Morales's menagerie, a cursory census of which counted, in addition to the turkeys and dogs, chickens, rabbits, hogs, ducks, and of course what Morales labels simply the *animales*, the variations on the theme of horse that are his work animals. Squash are key to feeding the menagerie, which acts as a caution to plant breeders.

Farmers here split the world of what might look like simply squash to the rest of us into two parts, the rough *chilacayote* and the true squash, which they call simply by the Spanish word for squash, *calabaza*. Of course, subsets of these do exist, but a simple approach to improving productivity from a plant breeder's point of view might be ignoring the *chilacayote* and breeding for varieties of the true squash with large, appealing fruit that will sell well in urban markets.

The livestock eat *chilacayote*. Some varieties of both kinds with smaller fruits still produce more forage per plant than those producing a few big fat squash. And then we come to the matter of seeds. A squash seed is about 30 percent protein and is rich in oils. In Mexico, people eat them. Like corn husks, squash seeds are an important market commodity. So, too, are squash blossoms, which go into cooking a special *quesadilla* served even in urban restaurants. So is squash candy.

Clearly, abandoning the milpa in favor of monocrop maize would simplify the farm according to the industrial model, but the impact would reach beyond the field to the rest of the operation. There wouldn't be as many animals, so not as much manure. But also lost would be proteins and oils for the farm family and for those families that depend on local markets. Simplifying also to a single marketable variety of squash would have many of the same effects. All of which argues for coupling farmers with plant breeders to let the farmers decide what sort of squash they would like to grow.

One way to measure the health of agriculture is to examine income. That's how it is normally done, profit and loss, but in Mexico this system would give agricultural economists a useless—if definite—number. You might even adjust to include the value of home consumption against the value of family labor. Maybe you could even adjust to include much of the income that goes unreported in Mexico, or accrues as corn husks. Robert Bye tells me that Mexico City alone has more than 250 open-air produce markets that move around the city, the evidence of the countryside's productivity in town, but that business is mostly off the books.

"Mexico runs on an underground economy. What's happening can't be measured," Bye says.

And even if you could account for all the money changing hands, how would you devise a system that accounts for the loss of squash blossom *quesadillas*, oils and protein from seeds, and the rush of vitamins and minerals from quelites against the store-bought lettuce and spinach that might replace them. This is about a broader definition of success, about a general well-being. You might begin to measure that by examining some hillside soil.

The Chalco Valley—and most anyplace, for that matter—is geographically different from the Sierra Norte de Puebla. The milpa project works in both, but Chalco offers a panorama—the quiet adobe villages spaced a few kilometers apart in flat, dusty grids of farm fields—easier to visualize as farming country. Here and there in Chalco stand the ruins of a big house, really the ruins of a social system that still leaves a deep mark on Mexican society. One such ruin can be found near José Morales's fields, a falling-down hacienda, and everyone around knows its name and history. Remnants of the feudal system the Spaniards imposed, and abolished, at least on paper, by the Mexican Revolution in 1917, these haciendas were dismantled only slowly into the middle part of this century. The colonial *hacendados* claimed the flat rolling volcanic valleys around Mexico City first, because they were the best and would raise wheat. The Indians kept the margins, places that look a lot less like farm country, and Sierra Norte de Puebla is nothing if not marginal.

A curious phenomenon of marginal landscapes arises, especially farther north in North America. Catastrophes such as floods and particularly glaciation swept the valleys and pushed flora and fauna to mountain ridges, where they weathered out catastrophe. In this way, they became centers of diversity. Like glaciers and floods, most human endeavor clings to valley floors. Against this, places like Sierra Norte de Puebla serve as *refugia*, that is, environments where a suite of plant life and a related way of human life survive. It endures here in a sense because nothing else will, so suited is it to existence at the ragged edge of possibility.

The place-name in Spanish means simply the mountains north of Puebla, a direct enough description, it seems, as we leave the main highway between Mexico City and the sprawling industrial city of Puebla, head north through another broad dusty valley where the drought has left miles of fields failed or fallow, then climb to foothills at the town of Oriental. The mountains here are the eastern leg of the Sierra Madre, the Sierra Madre Oriental, hence the town name. Our Volkswagen van climbs quickly into the trees, and the world phases from dust to lush—winding mountain roads, rain-forest peaks, steep canyons set with sharp-cut streams. Villages don't come into sight until one rounds a curve, and suddenly a few stone houses flash into view, then recede into forest again.

But not all forest. As the eye gets used to reading the shades of green, one realizes that most of the landscape is farmed in some fashion. The maize fields stand out most prominently, not simply because I am searching for maize, but because these look like no maize fields I have ever seen. They are nearly vertical, farms on land that people in my part of the Rocky Mountains would classify officially as "steep as a mule's face." Three-meter-high stalks of maize plants flower and thrive on barely walkable land, worked only with machete and hoe. Set among the rows, squash and beans signal only the beginning of the place's diversity.

One may get a sense of how far all this blossoms beyond maize by visiting the weekly market in Zacapoaxtla, a small city of 60,000 people at the south end of the Sierra. An Indian woman has laid out a patch of blanket a couple of yards long, the size of a supermarket checkout stand, on which she displays the produce of her land for the week. In this small space she has plums, black cherries, pears, chiles, prickly-pear cactus fruit, amaranth, *chilacayote* squash, radishes, lemon grass, thyme, mint, a local semiwild landrace of avocado, a popular local fruit called *mamay*, fava beans, sweet potato, and blackberry. All this probably comes from a plot of about an acre, the size of a suburban American lawn, a comparison not altogether irrelevant in that all Mexico, with its agriculture centered in maize, grows in total about 10 million acres of maize for its hungry masses. The United States grows a total of about 25 million acres of lawn.

Her blanket provides a window into a world of explosive diversity, a sense of which becomes clear only slowly through the monologue of Miguel Martínez Alfaro, an ethnobotanist with twenty-four years' fieldwork among the Nahuatl and Totonac Indians and mestizos of the Sierra Norte de Puebla. A tightly focused man, he has an apparently encyclopedic brain, the contents of which can be triggered to spill seemingly in toto with no more provocation than a single question. As we travel together for two days in the mountains, he catalogues for me every plant we see, complete with a cultural context of its uses and place in mountain society.

To begin with, the principal crops: maize, squash, beans, coffee, vanilla (sold almost exclusively for Coca-Cola), *pimienta* (or spice tree, sold as the key ingredient for Old Spice cologne), peach, apple, plum, pear, nuts, quince, cherry, and strawberry, marijuana, poppies, and on down a long list of plants and produce we have no name for nor vocabulary enough of taste, smell, and color to make them extend from a printed page.

There's a bottom line of sorts, which Martínez repeats to me a number of times: In two towns alone in this region researchers have identified 250 species of edible plants. In the whole region the locals use 350 species—both indigenous and exotic—for food. Another 300 have medicinal uses.

We have lunch in Zacapoaxtla at a small restaurant where platters of local specialties cost about two dollars. The same meal would fetch twenty dollars apiece in any American city, assuming one could find a restaurant that could make food so good. A pitcher of freshly squeezed tangerine juice accompanies the food. I try to explain to those at lunch, the Mexican scientists, the concept of fruit juice as it now exists in the United States, that freshly squeezed anything is mostly a memory and that we drink mostly flavored waters, the principal ingredient of which is corn syrup, such is our agriculture. I tell them the tangerine juice they drink like water is simply unavailable in most supermarkets and restaurants, and they look at me as if I am making it all up. I realize I am surrounded in these mountains by what anyone from the "developed" world would call poverty, and objectively, that is true enough. Yet the objective definition seems lacking.

The elevation of the Sierra Norte de Puebla rises from only 50 meters above sea level on the north, toward the Gulf of Mexico, to 2,700 at its crest, comprising everything from cloud forest to tropics, a mosaic of microclimates to which agriculture here must adapt. It does so, but throughout the region, with maize, beans, and squash at its center. Farmers on the tropical side add coffee to the mix but generally don't grow it in monocrop plantations. Instead, the typical farmer has maybe six or ten acres with coffee plants scattered among milpa fields. The coffee generally needs some sort of overstory to provide shade. It grows in a forest, overtopped by mahogany and cedar, which are occasionally cut for logs, then below that fruit trees such as citrus, but also macadamia nuts, *pimienta*, and vanilla. Beneath it all typically struts a flock of turkeys, with a few barefoot kids shooing them along, if only to get a better look at the flock of visitors combing their milpa fields.

For several days I have been bouncing along narrow dirt and rock roads winding to villages consisting of a few adobe houses. In villages, Spanish only gets us started, and conversations are translated from Nahuatl through Spanish to English. In some, whole schoolhouses empty out and black-eyed kids line the streets to see what we are about.

Fernando Castillo, Rafael Ortega, and I walk a maize field full of enormous stalks, each flush with foot-long ears. We've just talked with a farmer attached to a local organic-farming cooperative, and he has given us a lecture on preserving topsoil for future generations by recycling nutrients. Many of the farmers here are organized, especially the coffee farmers, who have formed a marketing collective and are attempting to sell organic coffee at a premium price. The organic farmer, who has a small but adequate house, is in his thirties, maybe, with a couple of kids. The importance of his small family is a fact I missed in the conversation, but it comes out later when we're in the maize field. Properly considered, it marks his success.

Engaging in mass selection of maize, Ortega and Castillo mark shorter plants that still produce healthy long ears. The end result of their work will not be dramatic. If it succeeds, it will improve farmers' incomes only marginally—but marginally counts for much.

"The difference between the poor and the less poor is great," Castillo tells me. In this region, an income of $2,000 a year makes a rich man. Many of the families that work these surrounding fields get by on a tenth of that. A good crop of maize can make the difference between $200 and $2,000.

The link to conservation can be seen at every schoolhouse we pass. Statistically the difference between the poor and less poor is not just money but the amount of children. The poorest of these houses, too small for a garage in the United States, typically holds more than fifteen people. As soon as income rises into the less poor category, the number of children per family drops precipitously, a relationship that holds worldwide. As a rule of thumb, more income generates smaller family size for a variety of reasons. People not desperately poor have time to educate themselves about birth control and the money to practice it. But mostly, they have access to resources other than their reproductive power and the free labor it generates.

Castillo points from the healthy field of maize to an adjacent plot now no longer a plot but a course of jagged rock. This is what happens when the poorest of these farmers works the land to its limits, not letting it rest, not letting organic content build, not maintaining cover crops. The farms of the Sierra Norte de Puebla are near vertical and will erode off the face of a mountain in a single sliding mass. Then the poor become the desperate.

The poorest farmers typically have five acres of land and cultivate all of it. The less poor, ten or fifteen acres and cultivate maybe four at a time, rotating and resting the remainder, planting trees on the remainder, working within the limits of the land. Here biodiversity becomes more than a buzzword.

"The people who have more income have more plants," says Castillo.


Just outside Mexico City lies the flat, polluted, and vacant plain of the old bed of dried Lake Texcoco, once the center of Aztec agriculture but drained by Spanish canals. And just beyond lies the complex that houses the International Center for the Improvement of Wheat and Maize (CIMMYT), the institution that grew from Norman Borlaug's work.

Bent Skovmand, a plant breeder, shows me a freezer that contains, arguably, the most important record of human culture extant: seeds of maize and of wheat, the largest such collection in the world. There are 19,000 landraces of maize from the Andes and Mexico and 145,000 samples of wheat. It is not a museum of our past, though, so much as our platform for a future. Outside the building, Skovmand has shown me a field of experimental wheat he has bred using a landrace unusually resistant to drought crossed with modern, high-yielding varieties. He argues that a single gene bred into seed in this way can be worth $18 to $20 million a year in increased yields, and it costs about 35 cents a year to store a wheat sample. They'll keep for a hundred years in the subzero temperatures of the freezer.

This collection started in the 1940s, because plant breeders even then were conscious of the idea of a genetic resource. With the prospect of a global climate more uncertain than anything we have faced so far, the idea of a genetic resource is gaining resonance.

"You never know what gene you are going to need until you need it," says Skovmand.

CIMMYT would seem to have that need covered, and if it doesn't, similar, albeit smaller, collections exist around the world. If it was a grain crop and ever grew, its seed is somewhere in storage, and can be stored for a very long time, much more cheaply than one can maintain the ways of life that made the genetic resource in the first place. The freezer and all those like it would seem to argue against the need for *in situ* conservation of genetic resources, which is on the surface the whole foundation of the milpa project. If so, this would not explain why CIMMYT itself is becoming more involved with *in situ* genetic conservation, is hiring ecologists and anthropologists along with the plant breeders, and is adopting the dynamic approach to conservation that guides the milpa project.

How else can we frame the question of why conservation of the milpa is necessary? The milpa project does the bulk of its work in the Chalco Valley, where the milpa system is vestigial, where farming looks more like the monocrop system that has taken the rest of the world. Chalco lies on the edge of change, but the system is far less threatened and far more intact in Sierra Norte de Puebla. We can credit this at first to isolation, and there's something to that, but the region still trades in global markets. The milpa survives here simply because nothing else will work. The land is marginal, and conditions are too harsh, too rocky and steep, for the flat monocultures of valley lands. So if conditions create a refugium for the milpa, why is it necessary to work for its conservation? Conditions will do that.

We think now of the erosion and the extreme poverty that can threaten the milpa even here, and the value for all of us that can be gained in studying what takes place on that fine line between the world's desperately poor and the poor. Much of the world's future will be decided along that border.

Beyond this, the work of the Sierra Norte de Puebla is not so much about conserving the milpa as it is about learning from it. We may not conserve genetic resources in a freezer, because genetic resources are not static artifacts, museum pieces. Genes are the living record of a dynamic process. The record cannot be preserved separate from the process. We do not so much need the creation as we need the creativity. What the milpa is today depends on all the human culture that has brought it to this point, depends on its constant communication with the teosinte, wild beans, and squash that breathes vital genes back and forth across the pervious boundary between wild and domestic. The milpa is endurance in the face of adversity, which is what we have come to learn from it.

# Roots

## Restoring Rural Wisdom

## [Peru]

There is a place where sustainability and biodiversity are more than buzzwords. It is a beautiful place, even on its surface, even before one knows its workings. It is a broad sweep of valleys and Andean peaks snowcapped year round despite being so near the equator. Valley floors run from 1,500 to 4,000 meters above sea level, the peaks nearly double that of elevations that would make this landscape marginal, even uninhabitable in most of the world. Here was cradled one of the world's great civilizations, if by great we mean one of the handful of seminal cords of agriculture that twisted together to support the world. South China's rice, the Middle East's wheat, Mexico's maize, and this region's potatoes.

By "great civilizations" do we mean wealthy? When the Spanish conquistador Francisco Pizarro first marched onto this plateau in 1532, he captured and held for ransom the Inca ruler Atahuallpa. Before Pizarro reneged on his word and killed Atahuallpa anyway, the Incas ponied up enough gold to fill a 22-by-17-foot room to a height of 8 feet. The gold was melted down and freighted back to Spain, a sufficient quantity to tip the economic balance of Europe. Were the Incas rich because the Andes happened to hold gold, or because their agriculture could produce the food surplus to support unproductive work like mining, refining, and smithing gold? At the time of conquest, this plateau supported an agriculture capable of feeding, it is estimated, 12 million people. Western eyes never witnessed it at its full height. By the time Pizarro arrived, the population had already been decimated by the smallpox that preceded him, leaving the mop-up work of conquest to guns and greed.

Wandering the Andes valleys and hills not long ago, a young American college student came upon some Inca ruins and was astonished: "Can you imagine being the guy who came over the hill and first discovered these?" Discovered? But they were never lost. Cajamarca, the city where Pizarro captured and held Atahuallpa, is still a city. So is Cuzco, the Inca capital. A colonial-period church starts near Cuzco's central square, ornate and European, yet its base is Inca stonework, the foundations of temples the Spanish razed, then built over.

A couple of blocks away one can walk Cuzco's open-air vegetable market, still the main source of produce for this city of 275,000, and see the foods that supported the Inca Empire. Potatoes, yes, *papas* in Spanish (the Inca's Quechua language gave the Spaniards the word), also tubers like *oca, ulluco, mashua,* roots like *arracacha,* grains like *quinoa* and *tarwi.* Also maize, fresh, fat kernels on fat ears, the kind of maize the Peruvians call *chocla* and eat like Americans eat sweet corn. *Coca* leaves. Better still to see this produce in a smaller market in a nearby town, where women spread small piles of maize and tubers on the ground on mantas—the carry-all shawl made of wool from sheep, llamas, or alpacas, dyed with plant stuff and woven on hand looms. The women sit stoically with rocklike faces in their tall-crowned, white straw hats in the pose that is the proper market etiquette of Andean commerce. It is up to the buyer to speak first.

The produce comes mostly from villages within sight, bigger plots on valley floors, smaller ones terraced up hillsides. To those whose imaginations of pre-Columbian conditions in the hemisphere are set in accounts of nomadic tribes in the northern plains, the Andes region comes as something of a rude jolt. Best to absorb its effects walking randomly down long valleys and over hills. Every few kilometers, one encounters another village, but also ruins. Every vista holds terraces and ruins, testimony to the density of the ancient human penetration in this place.

Again and again I wandered old terraces, stone walls ringing the hills with bases of old rocks piled by Incas who predated Atahuallpa. On top of these are new rocks and behind them freshly worked soil growing this year's crop of *mashua*, *oca*, and *papas*. Political discourse struggles to define sustainability, but these rock piles have been growing these crops for at least six hundred years, in some cases thousands. What more definition do we need?

We missed something in lumping the Incas together with the other great civilizations. Each of the other three—China, the Middle East, the Aztecs—rested on a cereal. The Incas had maize, but got it from the north; it was not the foundation of their agriculture. Tubers, especially potatoes, were. Furthermore, the seminal civilizations were, with the lone exception of the Incas', valley civilizations carved by low-elevation, broad river valleys, a flat platform, and a generally homogenous environment for a homogeneous crop. Images like amber waves of grain and fertile plains got strung together a long time ago. The Incas created agriculture on the edges, and most of the world's gains in agriculture in the next Green Revolution must be made on the edges.

The ride on a tourist train from Cuzco to Machu Picchu on the Andes' eastern Amazon headwaters slope advises that Inca life was exceptional. The last half hour of the ride on the shuttle bus from the train heads almost straight up, a long series of switchbacks that leaves a rain-forest valley floor to climb to the very peak of the mountain and the spectacular ruins of a mile-wide city. Above looms yet another mountain peak with sharp rock slopes jutting a thousand meters above the main city. Looking at that second peak, one can believe one's eyes are being deceived, but there at its very top one can see terraces for the agriculture that fed Machu Picchu.

This setting makes for a very different sort of agriculture. The various elevations, taken together with varying slopes and aspects, provide an almost infinite array of microclimates. To survive, the Incas needed to develop an array of plants and techniques to match those conditions: varying amounts of rainfall, earlier and later frosts, hard sun, less sun, high ultraviolet light. One crop and set of skills would not do. The situation demanded diversity, just as it demanded a developed system of trade to provide necessary goods that couldn't be grown in a given village.

How systematically had the Incas developed their techniques? Was it conscious invention or slow evolution driven by trial and error? That question resounds especially in another set of ruins, smaller and less famous than Machu Picchu, not far from Cuzco. The place is called El Moray.

Since it had been raining, the two-track dirt road to El Moray had turned to gumbo, and the group of us, a half-dozen honors students from Pennsylvania State University along with two instructors, Hector Flores and Marleni Ramírez, abandoned the bus to walk. Flores, a biochemist, tries to untangle the various chemical secrets of Andean foods. He and Ramírez, a couple, are the American half of the McKnight project in Peru. They were living in Pennsylvania, but both are actually native to Peru, where they did their undergraduate work.

The walk turned out to be longer than Flores had remembered, maybe 8 kilometers. We walked the crown of a broad, easy, green plain 3,500 meters above sea level. A sweep of hill defined one horizon. A rack of peaks fronted by the 5,700-meter El Chicón guarded the other. Now and again the fallow fields where donkeys, sheep, and cattle grazed were broken by neatly tended fields. We asked the few farmers directions and walked on through the afternoon.

El Moray is set in a sinkhole that appears abruptly in the plain, a limestone fault that drops sharply to the earth's floor maybe 500 meters. Approaching it is like walking to the edge of a deep lake that has been drained, leaving an inverted cone for a hole. The cone is ringed with terraces, seven concentric circles spread across its gently sloped bottom, then seven more above ring the cone's steep walls; then above that at least eight more semicircles form a sort of amphitheater along one elliptical lip of the hole at its top. The terraces toward the top are mostly in the disarray of ruins. By walking them carefully and taking advantage of the slanting light through the grass reclaiming the stone, the visitor can still see subtle patterns: two old water courses cutting the series of terraces down from the rim—irrigation ditches probably, from a reservoir above.

The bottom terraces, however, are not ruins but the familiar pattern of new rocks on old. A crop of fava beans ripens within.

Flores brought us here to consider a hypothesis set forth by archaeologists who have studied El Moray. They have reason to believe the Incas used it not as an ordinary farm but as an agricultural research station. It would be valuable as such not only because its enclosed slopes protect it from the more radical shifts of climate above but because its conical shape gives it slopes of all possible aspects and exposure to sunlight over a quick series of elevations. It repeats in microcosm the variations in topography over the range of Inca territory, an ideal setup for refining the knowledge the Inca surely would have needed to successfully inhabit this demanding, limited terrain.

To oversimplify a bit, the project in Peru is largely based in three villages and on three native Andean tubers the world has never heard of: *oca*, *mashua*, and *ulluco*. What relevance do these have for the rest of the world's agriculture? To begin with, the world may well hear of them someday. Part of the world's food future will no doubt include forgotten or neglected crops. Food writer Sophie Coe points

out that widespread famine in Europe disappeared after the introduction of New World crops a few centuries ago. She speculates that
this phenomenon may repeat:

> Who knows how many cultivars and entire edible
> species have been lost to us because of some decisions
> made in the seventeenth century for reasons that would
> be utterly irrelevant and ridiculous today. Reading
> about the foodstuffs of the Aztec, Maya and Inca and
> their European contemporaries, we come across many
> things which we either do not think of eating today or
> have simply never heard of. Given modern agricultural
> and transportation technology, plus the infinitely per-
> suasive power of modern advertising, perhaps we still
> have a chance to aim for a healthier diversity of food-
> stuffs, instead of eating more and more of the same fat,
> soft, sweet things.

Yes, the work in Peru may focus on tapping forgotten crops for
the world's diet, reason enough to undertake it. But put that aside.
Mostly the work aims at understanding how these crops fit in the
Andes and how to protect this unique and highly evolved form of
agriculture against the onslaught of global markets and the Green
Revolution monoculture. Interest in the area began building after
the 1992 Earth Summit in Rio de Janiero, with its emphasis on preserving biodiversity. About that time, researchers were recognizing
that Andean farmers had preserved a whole long list of food crops
unique to the region. They held a genetic resource that needed to be
conserved for exactly the reasons we saw at work with potatoes in
Chile. A wild potato relative from this region formed the base of that
work. Recognizing the value of the diversity, the Swiss Agency for
Development and Cooperation approached the International Center
for Potatoes at Lima shortly after the Earth Summit to begin re-
searching conservation.

The motive of preserving biodiversity interlocks, at the same time, with the goal of maintaining a decent diet for the people who rely on Andean agriculture. The process is not about museum conservation and, as was the case with the milpa system in Mexico, is under the same erosional pressures. Nevertheless, it has a more fundamental goal of simple understanding as well. The system has endured in a place where farming is not easy. To survive, the plants themselves have evolved a set of tricks, as have the people. The plants' tricks and the people's tricks may not translate directly to the rest of the world, but the very complexity and the methods of negotiating sustainability through complexity is something the rest of the world needs to understand. Mostly because the world as a whole is becoming a place where agriculture is not easy.

Complexity has existed in all these countries, but Peru helps us make the case for the importance of coming to grips with this protean topic. In some senses the world uses monoculture agriculture not so much because it works but because it is simple enough to describe. It lends itself to questions that can be reduced to fit within the space of a single dissertation or grant application. We can have some sympathy for that bias as we try to come to grips with a more complex system, facing—both as researchers and journalists—what is a many-faceted elephant and forced to render blind descriptions, as in the parable of old, by grasping only one or two limbs.


Besides Flores's and Ramírez's work at Penn State, the McKnight project in Peru centers on laboratories at the University of San Marcos in Lima, a research station in Cuzco, and three villages nearby. The villages—Picol, Matinga, and Q'eccayoc—are in walking distance of each other, and walking seems in keeping with the spirit of the place. They are small settlements of mud-brick houses, dirt streets, muddy yards full of serious-looking, barefoot kids and their dogs. The houses sit on the second-flattest ground. The very flattest—the best farmland—holds a soccer field. On a fine fall afternoon most of the adults are scattered outside the village along adjacent hillsides tending fields. A dozen or so men are working together turning fresh ground, fields left fallow for several years but now being readied for crops. They work in teams of three. Two men hold a wide, double-handled shovel that takes a meter-long bite of earth. They force-feed the tool by jumping and landing on it in unison to drive the point through the sod, then rock back to flip up the edge of the sod, which the third man grips and turns. They have no tractors.

Fallow land makes up an integral part of the farming system that spreads all over the hills. Toward the top of a long bowl, a patchwork of fields is laid out, each the size of a back-yard garden. The land is communally held, and some of it is farmed that way, but individual farmers also have their own plots. Sitting at the base of the array, a woman works at a hand loom weaving a manta in the morning sun. The fields grow potatoes, *oca*, and *mashua*, as well as tall blue flowers, maize intermingled with the native grains—quinoa and amaranth—and fava beans. A mud-brick wall protects one field from donkeys and such. Along its top the farmers have planted a row of cactus, making a living barbed-wire fence.

None of this scene is randomly arranged. A temporal rationale orders the fields, following a strict rotation. In the first season, a given plot will grow potatoes, the primary crop. The second year, it grows a mixture of native tubers intercropped with grains—barley especially—and with beans and the blue flowers, a species of lupine, a widespread wildflower. The same plant is native to the Northern Rockies and survives on my land there. In Peru, lupine is grown for the seed, called *tarwi*, which is dried, ground, soaked, and mixed with cheese to make a paste. Like all legumes, it acts as a nitrogen fixer. Following the second year, the field lies fallow, typically for three years.

Given this rotation, researchers must broaden their work to consider crops like the lupine, a key contributor to fertility, then broaden it further still to examine how it all flows through markets and trade.

What happens to *mashua* production, for example, when there's no demand for *tarwi*?

The immediate evidence of this whole system's dependence on trade is the village's cluster of greenhouses made of plastic and steel reinforcing rod frames. Each farmer who owns one has given us a tour, so we've seen maybe a dozen identical greenhouses. They are new and a source of considerable pride here, a direct result of the project's work. Part of the secret of sustaining Andean agriculture has been compensating for varied local conditions by trading germ plasm. Villages are constantly bringing in new varieties of tubers that may give them some advantage. Part of Ramírez's work is to study this systematic and probably ancient means of exchange, to find out why the farmers think it is necessary. They tell her they get new seed because they believe the old seed "gets tired."

Whatever the advantages or rationale, it is tubers that are exchanged, tubers that rely on vegetative propagation. Trading incidentally also includes fungi, insects, and viruses. That's what the greenhouses are all about. Instead of planting tubers directly into the fields, the researchers have developed a method of allowing them to sprout aboveground, then trimming and planting the disease-free sprouts and destroying the diseased tubers that made them. The farmers start the sprouts in the greenhouses, then transplant them to the field.

To a scientific purist, these greenhouses don't look at all like research, other than providing an answer to a simple question: Do they work? They look more like a Peace Corps–style effort at aid. As such, they offer an example of participatory research, a sort of hybrid between aid and research. The approach is controversial among purists to the point that some are loath to call it research. On one level it isn't, but is instead simply a pragmatic, seat-of-the-pants means of developing appropriate technology. On another level, though, the work is an acknowledgment of complexity. There is simply no way to isolate and examine questions, so the alternative is to take small experimental actions and monitor the results in actual conditions. So is the goal in these villages to help the farmers combat disease? Not really. That happens, but mostly researchers are here to figure out how all this comes together.

For instance, what about oca and its bitterness? A farmer has dug up a few oca plants for us. The tubers are attractive fat little fingers the size of small carrots. They range from orange to white, bitter to sweet, depending on the amount of oxalic acid present. High acid content means bitter oca. Flores has shown that oxalic acid confers resistance to weevils that infect the roots of the oca, just as tannin fends off birds from sorghum—but with a refinement: oxalic acid content is much higher in the peels of the oca, precisely where the plant needs armor against weevils. As with the tannin in sorghum, though, the farmers of the Andes have learned to circumvent the bitterness simply by peeling the tubers or by cooking methods, meaning the whole question of insect resistance is embedded in local knowledge of the plants and their cultivation.

*Tarwi*, the seed of lupine, is laced with all sorts of unpleasant alkaloids, which are removed by soaking the seed before processing. The nutritional value and palatability of some tubers are enhanced greatly by a unique drying process that takes advantage of the high altitude and fall harvest with frost in the air. For centuries, the Incas have freeze-dried tubers on frosty nights, resulting in a product they call *chuño*.

Researchers noticed that the farmers like to grow a little *mashua* with the oca crop. When they asked the farmers about this, the answers suggested that the farmers understood that the practice gave them some advantage, but didn't say what. The researchers found out and then demonstrated that *mashua* has definite antifungal and antibacterial properties. That's the case with many practices—they have long been part of the culture, but it's not clear how much the underlying rationale of the current culture understands: "We grow it because we like it," "I like the color," "It's better," "The seed gets tired."

What results, then, is a series of relationships between people, crops, and culture built on diversity. The connections are subtle. Rationale can get lost in translation between the frame of reference of the Incas and the equally limited frame of science. Two tubers might both contain starch and protein, but one has a vitamin the other lacks, or the amino acids that make up the proteins are different or keep better or complement each other. Starches have a different chemical composition that confers an advantage. Finally, understanding all the interweavings is not so much the basis of a research question as it is appreciating the underpinnings of an enduring culture.

The researchers are designing greenhouses, so are they here to do extension work, to help the farmers? Or are they here to do research, to have the farmers help them by telling them what they already know? After a point, the question rests on meaningless distinctions.

The people of the village have whitewashed one shed-sized mud-brick room, which serves as a sort of lecture hall. The researchers and farmers meet here to trade information, but we've been invited to eat as guests. Platter on platter piled high with boiled potatoes appear, along with the local cheese, fresh boiled maize on the cob, and *cuys*—guinea pig—served roasted with the head still on. Guinea pigs were domesticated in the Andes and have served ever since as a primary source of animal protein for farm households. Slaughtered sheep, goats, and cattle are so valuable that they are generally sold, not eaten in the village. We eat the whole meal smothered in a peanut sauce, heads and all.

Only days later does this meal come back to me somewhat painfully, not in the usual traveler's sense, but during a discussion at a lab at Lima's San Marcos University. I am speaking with Rolando Estrada, who, along with Flores, headed the McKnight project in Peru, and Vidalina Hendric, a nutritionist. For the duration of her study, she and a group of researchers lived in the villages I had visited. Each researcher was paired with a consenting family for the study period, allowing them to live with 11 of 18 families in Picol and 18 of 102 in Matinga. Each researcher would spend a night with a family, then rise each morning and watch the women make breakfast and all the other meals through the day, logging, weighing, and counting ingredients. The researchers weighed and measured every adult and child in each family, especially the children. They found that the children were malnourished—69 percent of the children in Matinga, 78 percent of those in Picol—this in the villages where farmers had made a feast for a visiting journalist and some privileged American college students.

Is this the condition of an Eden? Perpetual malnourishment? Some people think so. For instance, the writer William MacLeish, in his book *The Day Before America*, builds a case on extensive human fossil and archaeological records from around the world to conclude that agriculture was never the blessing we paint it as, especially not for the poor. Human skeletal remains from agricultural cities, both in the New World and the Old, compared with those of hunter-gatherers, show remarkable differences: the farmers were more disease-ridden, smaller, with more evidence of tooth decay and malnutrition. In part this was because the diet of agricultural people usually rested on only a few crops and a grain base. In part, the constant motion and the resulting diverse diet and living habits of the hunter-gatherers made for a healthier people. It is also true that hierarchy and status arose with agricultural societies, giving them the social tools that organized the stoop labor that got the crops grown and the temples built. The meat and the status foods like chocolate—the variety and quality—went to the rich few, then and now.

Even a central myth of Western culture sets agriculture in ambivalent light. When Adam and Eve fell from a state of grace, they were banished from Paradise to an agricultural existence. We may sing about getting ourselves back to the Garden, but Eden was something other than a garden. Indeed, some believe this story of Genesis arose and survived in ancient Hebrew tradition to encode a central tension in that society, a nomadic pastoral people falling from grace to become farmers. This same theme echoes in the story of Cain and Abel, a herder who killed a farmer.

If a degraded human condition was the lot of agricultural societies generally, however, a case can be made that it was less so in the Andes. This seems something of a paradox, considering that Andean civilization arose in what anywhere else would be regarded as marginal conditions. But the relative well-being of Incas was a direct result of those harsh conditions. Hector Flores once argued in a paper:

> The solution to this challenging geography was twofold. On the one hand, the rotation, soil conservation practices and irrigation canals which remain to this day a feat of agricultural engineering. The second part of the solution was biological, taking advantage of widespread adaptation of plant species to environmental extremes, namely the development of underground storage organs. Andean farmers domesticated a wider diversity of root and tuber crops than in any other agricultural system.

Also, the Incas seemed to avoid the worst abuses of hierarchy. Where the Aztecs and most Old World agricultural societies thrived on tribute from the peasantry, the Incas would periodically empty storehouses to feed the poor. Theirs was a more egalitarian society.

So why are the children of Picol and Matinga malnourished today? The full results of Vidalina Hendric's research makes this condition even more puzzling: the weighing and sampling of the villagers' diets found that everything they need for food security—ample quality and quantity of food that should give all those children normal growth rates—is right there in the villages. What they mostly lack is knowledge of how to prepare, store, and proportion the available foods.

185

"When you look at the food they are eating day to day, all the elements are there. The way they are using it gives the imbalance. They have everything they need," Hendric says through a translator.

Was such a condition of ignorance the mark of Inca society? Not at all, the researchers say. Both Matinga and Picol are relatively new communities, a generation old. They were created through a series of land reforms in 1969. That's when Peru finally broke up the hacienda system imposed by the Spanish conquerors. Under that system, farmers became landless peasants; subsistence agriculture gave way to cash-crop farming. The peasants did not grow traditional crops so much as they grew cash crops the landlords wanted. Slowly, the knowledge of the traditional crops faded. Land reform then divided the land among people who had been peasants for all the generations since conquest. Vestiges of traditional agriculture did survive, but in varying degrees. The newer communities cobbled together by land reform drew people who had forgotten essential pieces of the cultural weave. Some didn't know how to freeze-dry potatoes. Some didn't know how to make cheese. So far, there has been no systematic attempt to recover the skills.

Communities still exist where much of the traditional knowledge remains intact, but those are isolated and resistant to incursions by outsiders, particularly outsiders from universities. Estrada agrees it would be much better if the researchers were able to do their work in the older communities, but they are simply not welcome. It was hard enough getting cooperation in the new villages, especially when agronomists set up field tests for pest resistance complete with unprotected controls.

"The farmers said, 'Why should we work with you when you feed all the crops to the pests?,' " says Estrada.

Over time, the researchers were able to establish a relationship with the three villages, and now a majority of the residents there participate in the work. But of necessity, the work becomes a mix of research, of learning what the villagers know, and of teaching them things they should know. One could make a case that this is not pure science, but the case quickly evaporates in the presence of malnourished children.

Virgilio Rael Piñeda, a futurist who heads the Institute of the Future in Lima, serves as an adviser to Estrada. He is the sort of guy who quotes Alvin Toffler, the sort you'd expect to be touting fiber optics and high-speed data transmission. To some extent he does, but mostly he speaks with me about agriculture and poverty.

"Before the Second World War, we had poverty with food. Now we have poverty without food," he says.

Nothing in this dire set of affairs places Peru apart from most of the developing world. Up to now, the developed world has written the prescription, which has been development through industrialization and through world trade. Indeed, a big part of Peru's power structure is committed to exactly that, as the pockets of upscale malls and fast-food joints on the outskirts of Lima can attest. The government has also bought the prescription, which is why it ignores rural poverty and subsistence agriculture. The symptom of that neglect, a symptom present to varying degrees throughout the developing world, is a complete void in programs for teaching farmers.

"Ag extension doesn't work," Piñeda says. "The government could take this role, but it doesn't happen."

Those working in food security not just in Peru but worldwide tend to take government apathy, even antipathy, as a given. Instead, people like Piñeda and Estrada form coalitions with a string of international organizations, foundations, and institutions like the International Center for Potatoes and from this base cobble together an agricultural future. Piñeda would call that model postmodern, in that it depends not on the hierarchy of government and corporations but on a network of independent actors. At its heart lies the concept of self-sufficiency. Each village needs to produce what it requires, supplementing those needs by limited trading with other villages in the region, using established markets to swap germ plasm and altitude-restricted crops. He believes that if any place can pull this off, it is Peru, because Peru still has the base of diversity that built the Inca civilization, coupled with surviving traditional knowledge, in turn coupled with sophisticated technology.

"The only solution is to produce our food the way the Indian communities do now," he says. "We have the advantage that our agricultural systems have maintained biodiversity. Now we have to maintain both that and the modern. We have to make efforts to keep the actual cultural systems in relation with all the other technology."

Piñeda lays out an argument for what postindustrial designers would call a networked solution, autonomy and coalition replacing command structure. The idea is nice, but ultimately comes to rest on the ability to make coalitions, which is to say, rests on messy human dynamics. Just after my visit to Peru, a long-simmering feud between Flores and Estrada erupted into a full-blown battle. Things got ugly. After several rounds of attempting to mediate, McKnight's oversight committee voted to cut the whole program. Clearly the feud was the biggest reason, a fact neither trivial nor irrelevant to the larger task of shaping an agricultural future. Managing complexity means managing human dynamics—no small task. But underlying the failure was the complexity itself. That is, the program partly foundered on its own diversity; it was unable to pull its many elements together in a coherent direction—the same problem that cropped up in Zimbabwe, the other McKnight fatality.

It may have been because Peru was my last stop on this global journey, but I wanted to read more into the country's story than just the specifics of its immediate situation. Its diffuse nature and its basis in the ancient history of agriculture seemed to encompass so much of the larger task, to typify the challenge now facing the world. That sense was reinforced as I was leaving Lima and stopped to visit Rebecca Nelson, whom I had first met in Chile. Nelson is both blunt and analytical. On the matter of the differences between the Green Revolution and whatever it is we face currently, she has a solid perspective in that she has worked on both sides. She's a potato scientist now, specializing in late blight, the disease that caused the Irish potato famine. Potatoes are not a mainstream Green Revolution crop, but many people believe they will be the source of the biggest gains in the next Green Revolution. Managing late blight could instantly bring about huge gains in yields, but 150 years of fighting it with our best technology has ended in a draw. At first it appears as if Nelson's challenge is late blight, which is to say she is engaging a narrow technical problem, the very sort of work that characterized the Green Revolution.

But before joining this battle, Nelson worked with rice through CIP's cereal counterpart based in the Philippines, the International Rice Research Institute. The differences between that work and working on a crop like potatoes are huge, largely because of the homogeneity of the world's rice. The Green Revolution bred a handful of preferred rice varieties, working in isolation in a few test fields, and those varieties worked everywhere.

"We could sit in Los Baños [Philippines] and breed for the whole world," she says. "I could take a suitcase full of rice into a place like Vietnam and it would be enough variety to be multiplied to make seed for all the farmers in Vietnam." In a matter of a few weeks, she could set up field trials and multiplication that would effectively import the technology. End of story.

For potatoes, it takes three years "and a huge amount of money" just to begin the process. A single rice plant makes enough seed to plant 250 more; a potato, enough seed for only eight.

The Green Revolution relied on a highly exportable technology, but the Green Revolution is over. Future gains will come in crops like potatoes, and the going will be rough. The news is not all bad, especially if one considers the long-term potential of the Inca strategy. Potatoes and all these other tubers store their starch underground, out of sight, out of the world's mind until now, no small reason as to why we know so little about them. Behind all the theorizing, however, stands a simple fact that supported the Green Revolution: shorter plants put more energy into seed heads. There is only so much head weight a stalk will support. Potatoes face no such limitation. The plants don't have to support the stored starch. The ground does.

Simply by the management of late blight, the yield of potatoes can more than triple under some conditions. That's before any serious tinkering and breeding to alter the basic yield potential, the sort of tinkering that produced the cereals' big yield gains. That's the potential, but a complicated overlay of problems prevents reaching that potential. Nelson says 15 percent of Peru's potato fields—a country where people eat potatoes for breakfast, lunch, and dinner—are a total loss in most years. All the rest are heavily damaged from late blight.

"People get hysterical over the Irish potato famine. Well, hello, it's not like it's over," she says.

In the face of problems that persist despite the efforts, she shies from terms like Second Green Revolution. She says she wants to be a revolutionary, but the reality will be something closer to green evolution.

"All the Green Revolution was about making plants shorter. They just had this one easy-money proposition that they cashed right in on: making plants shorter. Sometimes you can cash in. Sometimes you can't. From there on, it's just real life."

# The Genie in the Genome

## Bioengineering in Context

Politics is the art of drawing lines and defending them. Science crosses lines, which helps explain the basic incompatibility of the two endeavors. We support ourselves with a collection of technologies derived from a science that is a numbing rush of expanding ideas; we are governed by a political world that traffics only in those ideas that can be reduced to fit on a bumper sticker.

Nothing that I can say in what follows will do very much to resolve or focus or even direct the growing public debate about genetically modified foods. That is, in fact, a defining factor of the nature of current public debate: nothing can be said to change it. It is a-rational. Not that I have any clear-cut answers to propose; it may well be too early for anyone to offer those. In fact, I was altogether surprised to find myself writing about the issue at all. I began this book largely ignorant of it, if anything inclined to side with those who counsel caution in rushing into this brave new world. I remember once thinking that given the relatively limited tools of breeding, a low-grade tinkering with the genome, human ingenuity gave us the fainting goat, the lop-eared rabbit, and, from the genes of a wolf, the Pekinese lapdog. If these examples do not give ample warning against messing with genes, I do not know what does.

I, like most, believed the issue of genetic modification of food crops had more to do with the likes of the Monsanto Corporation, hard-core industrial agriculture, monopolies, and making square tomatoes that fit in boxes—the hyperindustrialization of the food supply of the developed world. You who have joined me on this trip now know otherwise: genetic modification and genetically assisted breeding through markers is already ubiquitous in the agriculture of the developing world. Taken alone, this simple fact is significant enough to refocus the public debate. On a deeper level still, though, the technology is doing more than rearranging our relationship with the rest of life; it is helping to rewrite our understanding of life. This is not so much about invention as it is about knowledge. Simply put, what genetics has taught us has long since crossed or even erased the lines along which the public debate defined itself. That's the news, that's the fact that can be reported here. It may be too early to resolve the issue but it is not at all too early to begin a critical examination of the lines that define our discussions.

No matter what my opinion—no matter what anyone's opinion, even among the enlightened and privileged of Europe, Japan, or the United States—it isn't powerful enough to stop the technology. The genie is already out of the bottle, way out. Any attempts to stuff it back in will be about as successful as the Chinese have been in stopping use of the Internet. I worry more that by artificially drawing the lines of debate to focus on what is in reality an extremely narrow and ultimately trivial segment of the issue, we risk being blindsided by issues far more troublesome.

The political discussion necessarily rests on the line of definition that says genetically modified plants and animals are unnatural, that by tampering with the genomes of our food crops we have crossed a Rubicon of sorts. The name coined and used in Europe for such products is "Frankenfoods," drawn from our mythology about the perils of hubris as it applies to re-forming life. The lesson is that we ought not usurp the role of creator, and we must be prepared to pay heavy consequences if we do.

Indeed, there is some foundation for this fear. The public was roundly assured by scientists that pesticides were completely safe, especially the early, broad-spectrum pesticides now widely banned in the world. Much of the problem of those particularly nasty insecticides was their chemical basis. They were purely synthetic, based on a class of organic chemicals not found in natural conditions. That is, they created a basic chemical structure without precedent in the biome. The earth had not developed tools for cycling them, for digesting them, tearing them down, and sending their elements back through the chain. Nature is full of poisons, but poisons of its own devising that it can deal with. These it couldn't.

The more recent, less widespread, but more relevant example of genetically modified foods was mad-cow disease in Britain. The lethal disease that infected at least 215 people there stemmed from their eating beef infected with bovine spongiform encephalopathy. If lines of unnaturalness were crossed there, they lay in feeding animal by-products to cattle as a protein source. Left to their own devices, cattle are not normally cannibalistic. It's difficult to see how this situation is directly analogous to transgenics, but the social dynamics are. That is once again the scientific experts in this matter assured the public this practice was safe when it wasn't. Several commentators have pointed out the fact that it is no coincidence that a few years after the outbreak Britain is the center of resistance to transgenics.

We are only beginning to learn the hard, negative lesson of the 1950s and '60s: nature is an unimaginably complex system. We are not free to re-engineer the landscape in crude bounding steps, to introduce elements purely of our own design. A continuum of artifice may be available to us, but what seems to work best are those steps that go along with natural processes and cycles, nudging them in the direction they are going anyway. The further we veer from this course, the greater the peril.

So do transgenics represent a great variation from this course? The line alleged to have been crossed here lies in tampering with the genome, in cracking the central code of life and literally rewriting the software. Transgenic technology is a re-engineering of life, true enough, but so is conventional breeding. For at least ten thousand years humans have been engaged in selection, an artificial pressure on breeding populations. All the forms of life we call domestic have a genetic makeup, a code, that is artificial as a result of this pressure.

Since Gregor Mendel's work in the middle of the nineteenth century, our tampering has become increasingly systematic. In this century, plant breeding founded a real revolution. We have seen several examples of how extraordinary the steps can be in a simple breeding of plants. Hailu Tefera's tef project in Ethiopia could in no way be considered transgenic as the political debate currently uses the term. It is plant breeding. Yet recall the tedious work in emasculating a flower under a microscope to cause the normally self-pollinating plant to cross-pollinate. By no stretch of the imagination would this occur routinely in nature. Likewise, the potato project that began at Cornell University and now promises to produce an insecticide-free potato for Brazil and Chile is strictly straightforward plant breeding. Yet it crosses a potato with three sets of chromosomes with a wild relative with two. Natural?

The Green Revolution itself was based firmly on the development of hybrids, which in turn was based on crosses between varieties that would not cross unassisted. It stemmed from the crude and painstaking trial-and-error process of plant breeding: acres on acres upon years upon years of bumping genes together and seeing what resulted. In the end, enormous amounts of genetic material were transferred, big crude hunks of DNA. None of this was in any sense natural, yet it is as much a part of our world today as a Midwestern cornfield.

From this relatively routine breeding, a couple of human generations of even more extraordinary steps ensued. For instance, some interspecies crosses produced embryos that were not viable but were rescued with all sorts of artificial culturing techniques. New varieties have been created by directly fusing cells. Bacteria have been used to transfer genes, as have viruses. None of this work was controversial, or even widely reported. In the public imagination, the line was crossed only in the early 1970s, when scientists first made recombinant DNA, the basis of transgenics. That step, however, was but part of a long line of tinkering. A 1987 paper in the journal *Science* listed thirty crop species that had, through a variety of methods other than genetic modification, nonetheless received genes from another species to enhance their own traits. Eleven of those transfers came not only from other species but from other genera.

Not only our politics but our logic rests on drawing lines, on sorting the world into categories we understand: taxonomies, races, species, genera. And much of the revulsion toward transgenics is clearly based on the sacredness we impute to those lines. Once human societies held the line at different races. Many still do, but most of us have gotten over that. But species? Are we ready to accept a crossing of human and chimp, probably far less a leap than a crossing of teosinte and maize. One wonders how much of this inherent revulsion, this tribalism, shapes the reaction to transgenics.

Yet no matter how sound or logical our attachment to these lines, these divisions, nature clearly does not respect them. Genetic material flows from species to species with or without human help. To take an extreme example: The Pacific yew tree became well known in recent years for a trick it learned: producing a chemical called taxol that is useful in fighting breast cancer. Then researcher Gary Strobel at Montana State University discovered that a fungus growing on yew trees also produces taxol. He thinks the fungus learned the trick through a natural transfer of genes from tree to fungus—a natural transgenic not just across species and genera but across phyla.

Still another set of lines can be crossed. Even if we accept for a moment that our framework for sorting the world into species is not as solid as we think, that the walls between species are permeable, we still cling to the solidity of our uniqueness as individuals: I am who I am because of a unique genetic code and the way that code has reacted to a unique life history. This code is the postmodern version of the soul. Tinkering with it seems like a spooky experiment, a violation of the sacred, but the truth is, my code and everyone else's is a fluid bit of business, largely full of what scientists call "junk DNA," huge bits of useless information. The replication of its patterns, which happens every time a cell divides in my body, is messy and error-prone, so much so that a mutation occurs every few minutes. These copying errors are mostly inconsequential, because the information is mostly junk to begin with, but occasionally they cause some problems. Even more occasionally, they confer an advantage I pass to my progeny, a bit of creativity emerging from the sea of random noise.

Yet in a real sense, if I dwell too much on my central code, I miss a good deal of its essence. What I am on the genetic level is a wonderful mess of interactions with my environment, a chain of cause-and-effect expressions of DNA in response to conditions. Just as important, what I am in a real sense is not me. Remember that I have two sets of integral DNA in what can be called my genome—my own and that which is sort of my own but derives strictly from my mother's line, my mitochondrial DNA. Mitochondria, regarded now as an integral subset of each cell, have their own DNA because they were once separate organisms. At some point they fused to become a part of my cells, but they still come strictly from the egg side of my personal equation. So at what point did that fusing occur, or was it really a point, or do we just need to assign it a point in order to be able to think about what was once a separate species becoming us? What about all the microorganisms that I contain that have not crossed that arbitrary line? If you were to grind me up right now and set about the complicated business of assessing all my DNA, you would find that most of the DNA in my body is not me. It would belong to the range of microorganisms, the symbionts, that inhabit the shell along with me, digesting my food, conferring resistances, recording my life history. I couldn't get on without them. My body is really an ecosystem.

What, then, is an individual?


As I was ending my visit to Chile, Julio Kalazich wanted to make sure I took the time to see just one more field. We climbed in the back of a pickup truck and bounced out through some fallow land, down through a draw and some trees to a spot isolated inside a chain-link fence. It looked like a grain field, but the grain and fence existed only to meet regulations for isolation. Growing inside the ring of grain were transgenic potatoes. They'd been given genes from silkworms and chickens to express an antibiotic to a bacteria that infects potatoes.

Kalazich let go an embarrassed laugh as he showed the field and couldn't resist sharing an inside joke. The researchers have named this fenced-off field Punta Peuco. Punta Peuco is a prison in Chile, but a special sort of prison, one that in the United States would be called a "country-club prison," the sort reserved for white-collar criminals. In Chile they call it a five-star prison. The real Punta Peuco was constructed to hold a henchman of Pinochet's convicted of crimes during the junta. Only it doesn't hold him. It is an open secret in Chile that the prisoner simply leaves from time to time.


The philosophical distinctions we lay on genetics are not just academic. For instance, consider the work being done on India's chickpeas and Chile's and Brazil's potatoes. A primary goal of both programs is increased food security and reduced use of insecticide. In India the case for these goals is particularly acute, in that Green Revolution technology has pushed chickpeas off the land and protein out of the diet of some of the world's poorest people. In both cases, scientists are attempting to alter the plants' genetic structures to provide better inherent resistance to insects. In both cases, this sort of technology is particularly appropriate in an important sense, in that farmers need only adopt a new sort of seed or seed potato. The advantage is built in.

In the political world, however, the differences in the two programs will outweigh the similarities. All of us are permitted to cheer the trichomed potatoes because their resistance is the fruit of conventional plant breeding. The new potatoes get their resistance from a cross with a wild relative—a different species, but closely related. The genome of the resultant potato has been altered, but by a politically acceptable means. By the same right, however, we are to condemn the chickpeas, which acquired their resistance through the artificial transfer of a gene. It has crossed the line to "unnatural." The gene is from a different species.

Maybe. It would not be the first time we were roundly assured that a scientific advance was without a downside and it turned out otherwise, but what this debate misses is that there is no *maybe* with insecticides. They are certainly bad. So are we reduced to choices between the maybe and the certainly bad? No one ever said feeding a planet of 6 billion people would be without consequences.

The solution some propose is reducing genetic engineering to a personal choice by clearly labeling genetically modified organisms. What, after all, is more personal than one's choice of food? Ought we not begin, though, by labeling those potatoes that have seen a dozen or more applications of the mixture of pesticides the Chilean farmers call *la bomba*? Labeling soybeans that are grown by plowing up valuable wildlife habitat? Labeling tomatoes grown through exploitation of cheap immigrant labor? The focus on one set of distinctions loses others.

The researchers in Nanjing were attempting to deal with wheat scab, again through genetic modification, by introducing resistance genes from a wild relative. Never mind that conventionally bred wheat has benefited more than any other crop from the transfer of genetic material from wild relatives, all carried out through some highly unnatural breeding methods. The wheat from Nanjing is genetically modified, so it is suspect. What is not suspect is the way it, or any other wheat for that matter, is grown. By focusing on the genetic controversy, we miss the fact that the scab itself is a cultural issue, the result of a monoculture of cereals. It could well be controlled by crop rotation, diversification, and other readily available practices that Green Revolution thinking has undermined. The political debate condemns "genetically modified" while remaining dumb about a practice not only potentially but demonstrably many orders of magnitude more damaging to the environment.

Nearby in Shanghai, the crop is rice; the problem a stripe virus. Once again, the tool is genetic modification, but not of the rice, not even of the bug that carries the virus, but of the microorganism that lives inside the bug. So for those people who worry about the safety of genetically modified food, do we have cause to be concerned about eating this rice? Hard to imagine how there could be. It has not been altered in any way and probably would slip neatly under the radar of any bans on trade in genetically modified organisms. No controversy, no label. Which of course obscures the fact that this program crosses a line probably more significant than any other in the book. By and large, our tinkering, genetic and otherwise, has been concentrated on the world of domestic plants and animals. This particular Chinese program transgresses this limit by modifying a wild organism, something completely uncontrolled. It seems to me that this line ought to be crossed with more caution than all the others, but the focus of the debate dictates that the opposite is true. Remember that researcher Frank Richards said the engineering of symbionts is particularly attractive precisely because it skirts politics, forges the path of least resistance. It ought to be the path of greatest resistance. It is one thing to mess with the genes of a sheep in a pen, quite another to mess with those of a naturally reproducing population. If the sheep experiment goes wrong, it can be stopped. How do you recall an entire population of bugs?

I don't wish to make the case that genetic engineering is a peril-free technology, and here another important line needs to be drawn. The line can pop up quickly in a couple of parallel discussions. Opponents of genetic engineering tend to consider the American agribusiness giant Monsanto to be the devil incarnate. Partially this is because the firm has been a leader in developing the tool in straightforward on-farm applications, notably Roundup-ready soybeans, a genetically engineered variety of soybeans resistant to the herbicide Roundup, which allows farmers to control weeds in the soybeans without cultivation. This product is only the flagship enterprise, though, of a whole string of tinkering that has allowed the company to gobble up a range of competitors to become the giant in the field. Easy to see why Monsanto has become the focus of condemnation.

Yet I have also been in a room full of maybe one hundred researchers, all food scientists—most engaged in some sort of genetic work and thus called "gene jocks"—and heard Monsanto condemned every bit as roundly as in discussions among those opposed to the technology. One of those discussions was particularly illuminating in that the topic was intellectual property: the fact that genetically engineered organisms can be patented to permit only their "creator" to profit. The difference between this group of scientists and Monsanto's is that they are largely public sector scientists, academics, working to feed poor people. Specifically, they intend for their work to be distributed freely to the people who need it. All the scientists in the room were in some way associated with the McKnight program, which has an explicit policy stating that any results of research must remain within the public domain. Yet an expert in intellectual property was telling the group that to remain free, the work would need to be patented, restricted, such was the environment of biotechnology. If not patented, corporations like Monsanto can take (and have taken) an engineered organism, modify it slightly, and patent it.

Genetic engineering combined with the concept of intellectual property has made this possible. Indeed, Monsanto itself is a case study of how a revolutionary technology coupled with a legal system that doesn't fully understand its implications can create a monopoly.

Monsanto has grown in recent years by swallowing its opposition, by controlling patents like properties on a Monopoly board. Technology in part made this possible, a fact worth worrying about. In this worrying, however, a critical distinction must be made between public and private sector development of genetics. No technology is value neutral, so it's not the same as saying everything will be fine as long as only do-gooders handle the tool. Nonetheless, there is a difference, and the debate needs to acknowledge it.

Out of the blue one day I got a call from Curt Foreman, who works with the Fish and Wildlife Service in South Dakota. He wanted to speak with me, not about farming or about genetic engineering, but about some earlier work I did on the destruction of a vast ecosystem, the American grassland, by farming. That great prairie was plowed up and farmed. In places like Iowa, less than 1 percent of the native habitat remains. People generally think this destruction occurred during white settlement in the nineteenth century, a rapacious era of American industrialism, but most of it occurred in this century, the worst of it in this generation, well after the Dust Bowl told us we ought to know better. Foreman was calling to tell me it still isn't over. South Dakota has lost more than a million acres of grasslands to the plow during the past six years. The battle for wildlife habitat is being lost. The culprit is soybeans, which couldn't be grown in South Dakota before but can be now, and since they are profitable, grow they do. Are they Roundup-ready? As a matter of fact, they are, and so genetic engineering becomes a factor. But not the biggest. It also happens that soybeans are drought-tolerant. Drought tolerance, as we have seen, can just as easily come from conventional plant breeding. In fact, it is precisely the goal of much of the breeding happening today. In any case, the new varieties of soybeans will grow in the arid stretches of South Dakota, where soybeans wouldn't grow before.

Agriculture has more environmental impact than any other human enterprise. With it, we have altered living conditions on the globe. It created the population glut that must head any list of disastrous environmental impacts. The very makeup of the atmosphere reflects our use of nitrogen in chemical fertilizers; our agricultural pesticides register in every place touched by water. Irrigation has claimed most of our rivers; runoff of pesticides, fertilizers, and squandered soil foul what remains. People worry a lot about the environmental effects of genetically modified organisms. That's probably a good thing to worry about, just as it would be a good idea to worry about the environmental effects of the rest. It takes some stretch of the imagination to agree with the critics' charge that genetic modification could create an environmental catastrophe, but we know for sure that farming is already an environmental catastrophe. Basically, we need to see the forest, instead of focusing on the single prominent tree of genetic engineering.

I began this book with a trip to Madison, Wisconsin, to visit the labs of Robert Goodman, a plant pathologist who also chairs the committee of independent scientists that supervises the McKnight research. By the end of my year and a half of investigations, I had encountered Goodman a number of times, and had come to appreciate his thoughtfulness and reasoned responses. More than one developing-world scientist I met, fascinated with the nuances of the English language, thought Goodman's surname particularly apt. I had a reason for returning to visit him: the main of his career shoots right up the center of this issue. He took his Ph.D. at Cornell in 1973, working with plant viruses. Recombinant DNA did not explode on the scene until a couple of years later, but it turns out that a background in virology was a good platform for entering this new world. Viruses are nothing more than bits of nucleic acid: RNA and DNA. They are a chapter of the larger book that science was learning to read just then.

Goodman took a position on the faculty at the University of Illinois, but in 1981 he got a call from an old friend at the University of

California at Davis. Would Goodman be interested in joining a start-up venture being launched to tackle and develop what we've come to know as biotechnology? He would indeed. As vice president for research and development at Calgene, Goodman oversaw such landmarks in the field—and in the public controversy—as a tomato genetically engineered to maintain a sweet flavor. Eventually Monsanto bought Calgene. By then Goodman had left, for what he says were both personal and intellectual reasons.

Goodman's perspective on genetic engineering can help shed light on the debate. Recall the case I made about the difference between the research in India on chickpeas and that in Brazil and Chile on potatoes. I argued that there is fundamentally no difference, but Goodman would maintain there is. He sees more potential for lasting value from the potato work, but not only because the chickpea work is based in genetic engineering. India's work is based on a single gene, a narrow resistance from a single protease inhibitor. The pest, the pod borer, has already demonstrated it can evolve past this line of defense and has done so as many as six times. This is the trap.

"In thinking about genetically engineering plants," he says, "the same issues arise as arise in traditionally breeding plants. You are working with single genes with large effects, breeding major dominant genes. There is going to be a spectrum of responsiveness to those single genes over multiple generations, each generation selecting and continuing with an increasing number of individuals that have the ability to defeat the resistance gene. The effect of using single genes, whether transgenics or single dominant genes in a breeding program, is that you get on a kind of treadmill."

An alternative is available, however. More than a few species of plants have cut a long-term evolutionary bargain with a string of pathogens. The trichomed potatoes are a good example. The key difference is that resistance rests on a complex array of responses, multiple genes, multiple pathways, a chess game of cause and effect patiently worked out over the eons in farmers' fields, in research plots, and in 400 million years of plant evolution.

The problem with the current round of transgenics is that they focus heavily on the single gene, a limited tool. The gene from the bacterium *Bacillus thuringiensis*, Bt, which produces a "natural" insecticide, is a good example. In 1999, when controversy broke out over monarch butterflies and Bt corn, plant breeders were already shying away from Bt. Ward Tingey, the entomologist with the potato project at Cornell, said in an interview well before the story broke that breeders had already decided Bt was a one-trick pony easily overcome by pests. It didn't matter if it was bred into the line, a transgenic, or sprayed on, as organic gardeners do. The pests were going to beat it.

Even more of a devil can be found in the details. Monsanto and others had invested years of research and capital in what had become a lucrative business. Since 1996, American farmers had converted 40 percent of the nation's cornfields to genetically modified varieties. Monsanto was already firmly committed. Instead of using what molecular biologists call "tissue-specific" genes for Bt, which cause the chemical to show up only in certain parts of the plant, Monsanto chose to get more bang for its buck by expressing the chemical generally, not just in those portions eaten by pests. The result was Bt in the pollen, which is not a target of pests. It was the pollen that killed the butterflies.

If all this sounds an awful lot like the early days of broad-spectrum pesticides, it is because the thinking is still one-size-fits-all. These are the grounds on which one may object to genetic engineering of plants. Those who would play creator ought to at least try to learn something of the intricacies of creation.

Goodman says the result of the controversy is a sort of institutional inertia. The Monsantos of the business world are locked in a time warp. Because of their capital investment and the lag time from development to market—and the resulting need to make all that investment pay for itself before introducing new ideas that would render it obsolete—commercial development is running maybe a decade behind the actual science. It might as well be centuries, at the current pace of change.

"In the meantime, the genetic side of the science has developed dramatically with mapping, marker-assisted breeding, technology of sequencing the entire genome. Much of that has been done in the last five years. It is an order—no, orders—of magnitude in terms of the amount of information that can be generated. Because of the massiveness of it, it completely changes the way you think about asking a scientific question in biology," he says.

Goodman draws the line between simple transgenics, the insertion of single genes to exploit resistance along a single path, and what he calls the "genomic" approach. The latter uses our rapidly escalating ability to read the entire genome, the entire genetic code, and lets it guide action, whether conventionally breeding plants or transfering suites of genes from wild plants with unique tricks of resistance. It goes beyond altering the genome, from nature to nurture. That is, a close reading of the genome is showing genes for such traits as resistances that are never expressed for lack of triggering conditions in the environment. This raises the possibility of working more closely with plants to provide those triggers instead of rejiggering the DNA. Such research implies a deeper partnership with plants and the development of specific agricultural practices that trigger gene expression. The difference is like that between learning to make a clarinet blast a single note and arranging and conducting a score for a symphony. When it became clear that the biotechnology business was working on the former, Goodman opted out and left Calgene.

"The intellectual part of it [the decision] was real, of thinking about approaching agro-ecosystems as the complex genetic and ecological situations that they are, and trying to think about ways of using modern genetics and molecular methods which are so much more powerful that we can deal with more complex situations or phenomena than we could before. Generally the motive was to try to build a program that would deal with the complexity of the way plant populations and microbial populations actually function in nature.

"I have a pretty radical, maybe even foolishly naïve view in terms of what is going to happen. I think that the genomic approach is going to outstrip, in terms of impact on the commercial, anything that comes from transgene technology in a matter of a few years . . . I think we're going to look back at that in just a few years' time as being primitive and almost arcane," Goodman says.

Two weeks after I interviewed Goodman, a U.S. biotech firm announced an advance that illustrates his point. The firm had developed a method it calls "chimeraplasty," a means of altering genes at a specific site to stimulate a naturally occurring DNA repair system in the cells. The technology uses no "foreign" genes, but produces results that look the same.

Let us cast our minds back to where we began with a square foot of soil and its microbes. Scientists started truly seeing into this world for the first time by beginning to understand how to read DNA. Simply put, they could now see life where life had not been seen before, the microequivalent of Columbus finding the New World. In this light, genomic wisdom extends well beyond parlor tricks like sweeter tomatoes. It is knowledge. Goodman and his collaborator Jo Handelsman developed a concept they call the metagenome, which is not a genome of an individual or even of a species but a total genome of all the organisms in a system, the complex array of information acting and reacting with other pieces of information. We think of ecosystems as complex interactions among species, but more fundamentally, they are even more complex interactions among suites of genes. Goodman thinks we're heading toward a thorough reading of this more fundamental scale, the reading of a system's metagenome.

Probably the political world will find none of this comforting or, for that matter, even comprehensible. Remember, it took decades for Columbus's discovery to be known around Europe, partly because mapmakers had invested money in the old maps and wanted the stock sold before they printed new ones. A new view of the world takes time to absorb. Nonetheless, there seems to be no way to completely unlearn what has been learned, or to unthink what people like Goodman have thought.

One of the things he thinks is that the current debate is dangerous, because it is so narrowly focused on transgenics. By stirring up public reaction it threatens to obstruct the wider path of scientific investigation.

"There is the danger of building a hostility toward genetic technology of all kinds," he says. "That, I think, is an untenable political environment for the future of food production." All agriculture, from the crudest process of selection through the Green Revolution, has been based on redesigning the genetic structure of plants.

"At a fundamental level, genetic improvement, down to domestication of a modest number of plants species as our food supply, is integral to human society," Goodman says.

# A Common Ground

## Food, Cities, and the Integrity of Rural Life

The work of Prabhu Pingali, an economist and a grain guy, is anchored at Mexico City's CIMMYT, which in turn anchors it solidly in the Green Revolution. The center is the institutionalization of Norman Borlaug's early work. A continuum of views exists among food security experts. One of its poles is represented by the belief that the solution to more people is more grain produced through industrial agriculture methods such as irrigation, more fertilizers, large-scale commercial farms, and improved (meaning higher-yielding) varieties, the old-guard Green Revolutionaries. Place Pingali near that pole. But just near it. Very few unrepentant, pure-form Green Revolutionaries remain. The notion of a Green Revolution is a mental construct, a simplification born of a second simplification: that the task of agriculture is growing food. On its most fundamental level, the Green Revolution made the basic calculation that every human needs 2,200 calories a day, then set about developing an equation that harmonized that number with the number of breathing bodies on the planet. If we have learned anything, it is that the job of agriculture is not simply feeding 6 billion today, and however many X billion a decade hence. That's not really what this is about, nor is it what this book has been about. Seeing the issue as only a simple matter will get us in terrible trouble. Old-line Green Revolutionaries understand this fact, as do most others working in the field.

Pingali stands before an audience of food scientists explaining a slide projected on a screen behind him, two simple columns. The first lists megacities, those cities in Asia that each contain a population of more than 10 million. The second column contains the list of cities expected to become megacities in Asia early in this century. There are twenty, a doubling in less than a decade. Then Pingali singles out one in the second column, a newcomer: Hyderabad, India. Ever hear of Hyderabad, India? Pingali has. It's his hometown. There were 300,000 people in it when this middle-aged man was born. To put this in the context of the American experience, which has no precedent for this sort of growth, imagine if you told the members of today's graduating class from a high school in Colorado Springs, Colorado, that their city would hold 10 million within their lifetime.

We tend to identify raw population growth as the problem, but something beyond population growth is driving the explosion of these cities. The total mass is growing, true enough, but at an even greater rate, the mass is shifting, coagulating in dense pockets of population that must be fed by dense packages of carbohydrates—cheaply, because most of them are very poor. What they will get is not much more than the daily dose of gruel based in the primary grain of the region. If they are lucky. To feed them, commodity prices must be kept low, which means prices to farmers must be low, which means that the other great collection of the world's poor, the rural poor, will be increasingly marginalized and impoverished to make room for efficiency.

A feedback loop drives this problem. Large, impoverished urban populations demand cheap carbohydrates, which in turn demand large-scale industrial agriculture. The world has survived to this point by depending on the methods that the Green Revolution gave us, but almost everyone involved thinks those methods are not sustainable. That's one problem. The second is worse. Large-scale in-

# World's Largest Megalopolises, 1990 and 2005
## Population (Millions)

### 1990

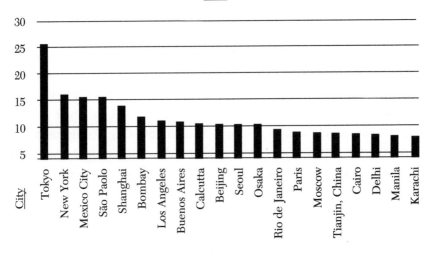

### 2005

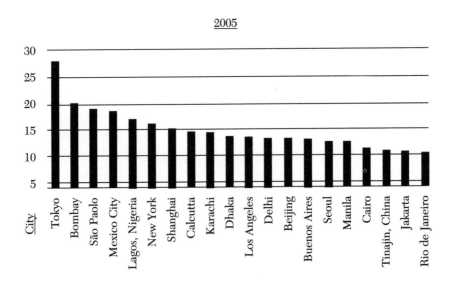

*Source: United Nations*

dustrial agriculture has been the major force in pushing people off the land and toward those megacities that grow far faster than the overall population growth.

The farmer-poet Wendell Berry cast the root question best as one of efficiency, as set in the American experience. More than half the U.S. population lived on farms at the turn of the century, just as more than half the developing world's population now does. Today less than 1 percent of the United States' population are farmers. We are told that technology's efficiency made farmers superfluous, so they moved off the land. The process continued in cities, where technology made human labor superfluous. In the United States we can ignore the effects by simply ignoring the urban poor who remain and the masses of unemployed in the megacities of the developing world. So if people are no longer needed on the farms and therefore sent to cities, where they sit idle in great sprawls of tin shacks, how does the system of technology serve humanity as a whole? That was Berry's question in a short and vital essay entitled "What Are People For?" That is as good a way of asking the central question as any I can devise.

The great demographic shift in the United States is generally understood by agricultural economists as largely a matter of economies of scale. That is, the capital-intensive methods of industrial agriculture dictate an ever-larger scale, generating the efficiency that makes farmers produce grain crops at surplus levels. This is an article of faith, but assume for a moment that it's true and understand that the U.S. experience is the driving model for the developing world. Places like India and China intend to make their agriculture more like farming in the United States and Europe. True, India has its megacities, but nonetheless, more than half its population remains rural. I asked a group of Indian scientists if they could imagine what their country would look like if only 1 percent of their people lived on the land. Their jaws dropped (literally) and their eyes grew wide. They shook

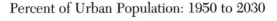

Percent of Urban Population: 1950 to 2030

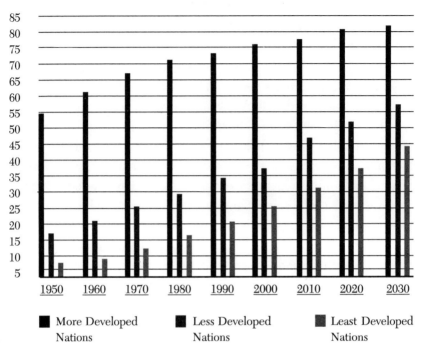

|  | More Developed Nations | | Less Developed Nations | | Least Developed Nations |

*Source: United Nations*

their heads but could not answer. Yet that's just what is happening. India is on the same path, as that growing list of megacities testifies. If development follows the U.S. model, then India's cities will be asked to absorb another half-billion people in the course of a few decades.

Reframe this question with another scene from India, one we have already visited. Recall the poor farmer with starving kids and a failed crop of chickpeas directly across the road from an experimental station's thriving fields. The genes in those robust chickpeas encoded a resistance to the very disease that was destroying the poor farmer's crop, and the seed that contained the genes were readily available in the area. This scene is a particularly stark example of an area agricultural experts call "extension." It is a telling word choice. Think of the extension of a line. Basically, the practice is generally drawn from a straightforward linear model that says researchers develop technology, then technology is "transferred" to farmers and everyone lives happily ever after. Only it doesn't work. Nowhere does it work well, and in most places it doesn't work at all.

That is not to say that technology does not get adopted. It does, but in Uganda, in India, and in Chile we saw clear cases of the most likely path of adoption. That is, a few generally successful, innovative farmers, early adopters, experiment with the new varieties and succeed. The bellwether strategy suggests that their neighbors will see their success and follow along, but more often it works a different way. The technology gives the early adopter a competitive advantage that he uses to drive his neighbors off the land. Those refugees then become the urban poor. I have no way of proving this, but my dark hunch suggests that this phenomenon was as much at work in depopulating rural America as was the inherent advantage of efficiency. That is, the failure of the family farm in the United States can really be seen as a failure of extension. Some farmers learned the new tricks and succeeded at the expense of their neighbors, who were, to use Woody Guthrie's term from the 1930s, "tractored out."

Never mind romantic notions about the importance of the yeoman farmer and the benefits of village life. Think of the untenable mass of cities. Can the world allow this failing to continue? The issue wrapped in this question for researchers, then, is one of accountability. There is no technology with neutral social effect, and powerful technologies can have powerful negative consequences. We understand this all too well in the environmental context, lessons learned with hard experiences like the near-extinction of the American bald eagle through farmers' use of DDT. We have learned the lesson well enough that we now (usually) ponder at length the negative environmental consequences of a technology before we turn it loose, as we currently are doing with genetic engineering. That same technology, however, has enormous social consequences, which go largely untested. The prime directive for those who would help the world's poor ought to be "First do no harm."

Toward the end of this journey, I laid all these issues on Robert Goodman's desk, unfairly in a sense, but also exactly where they belong. He is directing a $12 million program in ag research. That little bit of money will not reshape the world. It will, however, develop some technology, and who knows how that will ripple through the system. As the program has matured over six years, it has also become clear to Goodman that the old linear model (lab to extension expert to farmer) simply does not work, or works toward a different sort of end. Some would consider the U.S. model of extension the example the world should follow, with an extensive network of land-grant colleges begun in the Lincoln administration and a flowering of a network of agents working directly with American farmers. This is the backbone of the American system, but remember the demographics it has wrought. Meanwhile, the American extension system has declined in importance. Goodman has an explanation for this.

"Extension was an effective mechanism for industrializing our agriculture," he says. "In terms of the public agenda its job is done. Now the question is: What good is it? So what does it mean to have a world such as we have in North America and Western Europe? What does it mean to have a world in which only a small percentage of the total population knows anything about growing food? That puts civilization in an extremely vulnerable position."

Yes, in global terms, but what about the specifics? What should a foundation like McKnight or Rockefeller or Ford do in the face of this issue? Agricultural research has by definition come from researchers, in itself a cultural issue. The world is good at research and knows how to do it. Scientists can home in on scientific problems to create technology. Then they turn this technology loose in the belief that extension is not their job but that of the culture. They are easily blindsided by the attendant social issue, which keeps them on the path of least resistance. Research is the path of least resistance for all of us. Our culture knows how to develop technology. We are at a loss to explain how it filters into society—a much more difficult question.

If science is weak in this area, government is a complete failure. In each of the target countries of the McKnight program, the government runs an extension service. McKnight explicitly told the scientists doing the research to link up with those services to foster farmer adoption of appropriate technologies. In some countries—Uganda for instance—that has worked reasonably well, but in most it has not.

"Was it naïve to think that by asking them to get involved they would then build the linkages? Yes, it was naïve. It comes down to not actually building the linkages but making the linkages work," Goodman says.

Part of the solution to this comes from Uganda's example. It understood from the outset that research needs first to be cast in a social matrix. That is, the country consciously decided that the goal of its agricultural program was to assist subsistence farming, as opposed to commercial export crops. That policy framework went a long way toward developing technology appropriate to the farmers who actually needed it.

Still, the framework alone won't get the job done, and the world is short on people who know how to do the rest, especially the world's governments. Don Duvick, a career researcher with Pioneer Seed, told me in one of my first interviews for this project: "One of the things that has impressed me so much and that is so clear in so many developing countries is that the people who are governing the country are rotten through and through, and that's all there is to it. The laws are to enrich the people in power and nothing more."

In the months I traveled the world and talked to people in country after country, corruption was one of the things impressed upon me, too. They might not be so blunt about it, but most in the field would agree with Duvick. One of the basic assumptions of food security work is that government cannot be counted on for help and, in many cases, can be counted on for just the opposite.

This assumption has created the beginnings of a sort of ad hoc network solution. That is, researchers are increasingly linking up with nongovernmental organizations, NGOs, volunteer workers in the various countries, to do primary research and extension. Goodman cites this as a positive development, and one that will grow. There is an important difference, though, in the quality of this network as opposed to linear models based on government extension. The NGOs are far more likely to be committed to the primary goal of improving the life of the poor and the attention to subsistence agriculture that implies. Simply growing more food will not improve the lives of the poor; sometimes it makes them worse. The network approach can easily include those inclined to ask questions about the overall quality of rural life, and they can help prevent technology from degrading it.

Governments are easily seduced by the development model, which places them on the path followed by the United States. This castigation of government is not the same as arguing for private sector solutions, by which we generally mean business. In fact, in much of the world, government has failed precisely because it does not act in the public interest but instead goes into business for itself. We need to build the links that make sense, even to government and business, while never forgetting to act in the public interest, toward the common good.

Realizing that these networks are developing is important to us observers in that they represent an important evolutionary step for the social framework of the evolution of agriculture as a whole. I have set up the Green Revolution and its industrial methods as one pole in this argument, but have been somewhat lax in defining the other. In an evolved solution, definitions are necessarily lax, but the evolution of these networks paints some of the picture. So do the nine projects we visited. We have already witnessed genetic engineering at work and have seen the enormous potential for genetic work to reshape crops. Yet in some sense that aspect of the work is not so very different from the Green Revolution, with its emphasis on increased yields of dwarf varieties of cereals. The farmer receives a new seed. But seeds alone won't solve the problems.

The broader themes, the hallmarks of emerging agriculture, involve increasing complexity and diversity—not relying on seed alone but incorporating the power of genetics into a system with broad integrity. Solutions will vary with location. One size will not fit all. The array of crops will become more diverse, especially when drawn from the genetically stored wisdom of native plants and forgotten crops. Cultural practices will become increasingly important. Local information will drive the process. Farming will become more attentive to its broader environmental context, not only by degrading it less, but by tapping natural forces for assistance.

You may have noticed as we wandered the nine countries that many scientists were talking to farmers in real conversations, which is to say, listening, not just talking. Remember the highlands of Mexico. Remember the farmer selection trials where plant breeders worked not in experimental plots but in real farm fields, with farmers, a joint effort in maize breeding, an example of participatory research. Remember the highlands of Peru, a place that still carries an amazingly diverse and deeply adapted system of agriculture, how much of the information for running that system has been lost to farmers because of colonial land-use systems that made them peasants and how pockets of knowledge remain, tantalizingly, just out of the reach of science's ken. We know people have the knowledge, but we don't quite know how to learn from them.

All this suggests the real breakdown of the linear model. Information and knowledge will no longer flow from top to bottom but will originate in and reverberate through every part of the system. Information flows among researchers and farmers that in the end could have them working on a common ground, a common field of knowledge. It may be difficult to define what will replace Green Revolution methods, but this concept lies at its core.

The genetic engineering business is going to get all the headlines, but these simple matters are potentially far more earth-shaking.

## A Common Ground

What must happen, and to a degree is happening, in agriculture is also an information revolution. If there was a key mistake of the Green Revolution, it was in simplifying a system that is by its very nature complex.

Farming is not just growing food. It is not simply a tool we use to feed however many beings our social structure generates. The way we grow food determines our structure, makes our megacities, makes us who we are. Agriculture is culture, at bottom about the integrity of individual lives. Those lives gain their integrity and value when they are deeply embedded in a rich environment of information. This *is* about growing good food, but more important, it is about making good lives. We will fail if we attend to the former without considering the latter.

# Selected Bibliography

Aldridge, Susan. *The Thread of Life: The Story of Genes and Genetic Engineering.* Cambridge and New York: Cambridge University Press, 1996.

Berry, Wendell. *The Unsettling of America: Culture and Agriculture.* San Francisco: Sierra Club Books, 1997.

———. *What Are People For?* San Francisco: North Point Press, 1990.

Briggs, Philip. *Guide to Ethiopia.* Bucks, UK: Bradt Publications, 1998.

———. *Guide to Uganda.* Bucks, UK: Bradt Publications, 1996.

Brown, Lester R. *Who Will Feed China?: A Wake-up Call for a Small Planet.* New York and London: W. W. Norton & Company, 1995.

Brown, Lester R., et al. *State of the World 1996.* New York and London: W. W. Norton & Company, 1996.

———. *Vital Signs 1999.* New York and London, W. W. Norton & Company, 1999.

Coe, Sophie D. *America's First Cuisines.* Austin: University of Texas Press, 1994.

Diamond, Jared. *Guns, Germs, and Steel: The Fate of Human Societies.* New York and London: W. W. Norton & Company, 1997.

Drache, Hiram M. *Legacy of the Land: Agriculture's Story to the Present.* Danville, Illinois: Interstate Publishers, 1996.

Durning, Alan Thein. *How Much Is Enough: The Consumer Society and the Future of the Earth.* New York and London: W. W. Norton & Company, 1992.

Duvick, Donald N. "Biotechnology Is Compatible with Sustainable Agriculture." *Journal of Agriculture and Environmental Ethics,* 1995, Vol. 8, no. 2, pp. 112–25.

———. "Plant Breeding, an Evolutionary Concept." *Crop Science,* 1996, Vol. 36, no. 3, pp. 539–48.

———. "Responsible Agricultural Technology: Private Industry's Part." *Pro Rege,* June 1990, Vol. 26, no. 4, pp. 2–13.

Flores, Hector. "The Future of Radical Biology? Connecting Roots, People and Scientists." *Radical Biology: Advances and Perspectives on the Function of Plant Roots,* H. E. Flores, J. P. Lynch, and D. Eissenstadt, eds. Rockville, Maryland: American Society of Plant Physiologists, 1997.

"Genetically Modified Food: Food for Thought." *The Economist,* June 19, 1999, pp. 19–21.

Goodman, Robert M. "Bringing New Technology to Old World Agriculture." *Bio/Technology,* August 1985, Vol. 3, pp. 708–9.

Goodman, Robert M., et al. "Gene Transfer in Crop Improvement." *Science,* April 3, 1987, Vol. 236, pp. 48–54.

Handelsman, Jo, et al. "Molecular Biological Access to the Chemistry of Unknown Soil Microbes: A New Frontier for Natural Products." *Chemistry and Biology,* October 1998, Vol. 5, pp. 245–49.

Ibrahim, Youssef M. "AIDS Is Slashing Africa's Population, U.N. Survey Finds." *The New York Times,* October 28, 1998.

Jackson, Wes. *New Roots for Agriculture.* San Francisco: Friends of the Earth, 1980.

MacLeish, William H. *The Day Before America.* Boston and New York: Houghton Mifflin Company, 1994.

Mann, Charles. "Reseeding the Green Revolution." *Science,* August 22, 1997, Vol. 277, pp. 1038–43.

———. "Saving Sorghum by Foiling the Wicked Witchweed." *Science,* August 22, 1997, Vol. 277, p. 1090.

McKinley, James C., Jr. "Fueled by Drought and War, Starvation Returns to Sudan." *The New York Times,* July 24, 1998.

McNeil, Donald G., Jr. "AIDS Stalking Africa's Struggling Economies." *The New York Times,* November 15, 1998.

Mellor, John W. "What to Do About Africa: Closing the Last Chapter on U.S. Foreign Aid." *Choices,* Fourth Quarter, 1998, pp. 38–42.

Noble, John, et al. *Mexico.* Hawthorn, Australia: Lonely Planet Publications, 1998.

Pinstrup-Anderson, Per, Rajul Pandya-Lorch, and Mark Rosegrant. "The World Food Situation: Recent Developments, Emerging Issues and Long-Term Prospects." IFPRI, Washington, D.C., October 1997.

Reeves, T. G. *Sustainable Intensification of Agriculture.* Mexico, D.F.: CIMMYT, 1998.

Ruttan, Vernon W. *International Agricultural Research: Four Papers.* Minneapolis: University of Minnesota, February 1998.

———. "Population Growth, Environmental Change and Technical Innovation: Implications for Sustainable Growth in Agricultural Production." *The Impact of Population Growth on Well-Being in Developing Countries.* New York: Springer Verlag, 1996.

Soule, Judith D., and Jon K. Piper. *Farming in Nature's Image: An Ecological Approach to Agriculture.* Washington, D.C., and Covelo, California: Island Press, 1992.

Specter, Michael. "Zimbabwe's Descent into AIDS Abyss: Little Hope, Much Despair." *The New York Times*, August 6, 1998.

Storey, Robert, et al. *China*. 6th ed. Hawthorn, Australia: Lonely Planet Publications, 1998.

Strange, Marty. *Family Farming: A New Economic Vision*. Lincoln and London: University of Nebraska Press, 1989.

Swaney, Deanna. *Zimbabwe, Botswana & Namibia*. Hawthorn, Australia: Lonely Planet Publications, 1995.

Thomas, Bryn, et al. *India*. Hawthorn, Australia: Lonely Planet Publications, 1997.

Thrupp, Lori Ann. *Bittersweet Harvests for Global Supermarkets: Challenges in Latin America's Agricultural Export Boom*. New York: World Resources Institute, 1995.

Timmer, Peter, Walter Falcon, and Scott Pearson. "Introduction to Food Policy Analysis." *Food Policy Analysis*. Baltimore, Maryland: The Johns Hopkins University Press, 1983, pp. 3–18.

"Winning the Food Race." *Population Reports*. Baltimore, Maryland: The Johns Hopkins University School of Public Health, December 1997, Vol. 25, no. 4.

Woese, Carl R. "There Must Be a Prokaryote Somewhere: Microbiology's Search for Itself." *Microbiology Review*, 1994, Vol. 58, pp. 1–9.

Zuckerman, Larry. *The Potato: How the Humble Spud Rescued the Western World*. New York: North Point Press, 1998.

# Acknowledgments

The author Isabel Allende advises, "Copying one writer is plagiarism; copying many is research," and that bibliographies are "a bore." I don't dispute that, but I have chosen to acknowledge the wholesale plagiarism that helped guide this book in a bibliography. Unlike Allende, moreover, I am a journalist, and journalists plunder thoughts and information in person as well as from the printed page. A lot of people helped me to formulate the ideas expressed here.

Heading that list is my friend Rosamond Naylor, an economist at Stanford who also sits on the independent committee of scientists that runs the McKnight program. The idea for the book project was hers, and she pursued it by linking me with the foundation. Further, she lent support throughout, allowing me to draw freely on her vast scope of ideas and experience in international agriculture, not to mention her seemingly bottomless store of energy.

The foundation itself continued that support in a similar vein. Particularly, Michael O'Keefe, then executive vice president of the foundation, granted me editorial autonomy, then gave freely of his time for interviews that helped frame my ideas. The foundation's logistical support allowed me to range around the globe gathering these stories, and Sara Whitehead and Kathy Rysted deserve special thanks for help and advice in this department.

In the early going, Vern Ruttan and Donald Duvick of the oversight committee granted interviews and prepared long lists of sources and suggestions, and newer members continued to steer me through the network. Molly Kyle Jahn gave me a crash course in plant breeding and hooked me up with experts at Cornell University, such as Henry Munger. Alison Power, also at Cornell, supplied many key ideas about agriculture, especially as it relates to the larger environment. Prabhu Pingali helped me navigate through Mexico's CIMMYT and provided a sturdy sounding board for my ideas, especially when I disagreed with him. John Axtell invited me to Purdue, where he spent a long afternoon sharing his expertise on Africa and sorghum.

In addition to being able to interview these people individually, I was privileged to sit in on a number of oversight committee meetings, where I watched and listened to them work with international colleagues such as Jojah Koswara, Almiro Blumenschein, Usha Vijayraghavan, and Qifa Zhang. A great deal of credit for the open flow of ideas I observed—not to mention for the character of the project as a whole—goes to Robert Goodman, who chairs the oversight committee. Bob made time for me repeatedly throughout the project, freely sharing his wisdom and knowledge but never once attempting to limit or steer my inquiry.

During the course of my research, McKnight hired a team of independent evaluators—Ronnie Coffman, Rebecca Nelson, and Cynthia Bantilan—to assess each of the projects as well as the functioning of the whole program. I was given full access to their work, which became an important resource. I was lucky enough to travel in the field with Rebecca Nelson, and I particularly came to value her experience, her sharp, critical mind, and her generous nature. In the steady stream of E-mail between my home base in Montana and hers in Peru, she helped knock some of the nonsense out of my ideas and bring them back to the reality she knows so well. I am grateful for her wisdom, support, and friendship.

I am especially grateful to the grant recipients themselves, who took the time to show me their work and to help me with the logistics of travel in the developing world. The generosity was overwhelming, the experiences life-changing, and I thank them, both for their contributions to this book and, more important, for doing what they do.

Of course, I did not write a book so much as a manuscript. My editor at North Point Press, Rebecca Saletan, found the book in the manuscript. Dedicated, thorough editors are an endangered species in the publishing world, and I'm grateful that a bit of habitat survives to support Rebecca, just as I am grateful to my agent, Elizabeth Kaplan, for pointing the book her way.

Finally, and as always, my biggest debt is to my wife, Tracy, for keeping me going and understanding why I do what I do.